Fracture Mechanics 1

Fracture Mechanics 1

Analysis of Reliability
and
Quality Control

Ammar Grous

iSTE

WILEY

First published 2013 in Great Britain and the United States by ISTE Ltd and John Wiley & Sons, Inc.

ISTE Ltd
27-37 St George's Road
London SW19 4EU
UK

www.iste.co.uk

John Wiley & Sons, Inc.
111 River Street
Hoboken, NJ 07030
USA

www.wiley.com

Library of Congress Control Number: 2012950202

British Library Cataloguing-in-Publication Data
A CIP record for this book is available from the British Library
ISBN: 978-1-84821-440-8

MIX
Paper from responsible sources
FSC® C013604

Printed and bound in Great Britain by CPI Group (UK) Ltd., Croydon, Surrey CR0 4YY

Table of Contents

Preface

This book is intended for technicians, engineers, designers, students, and teachers working in the fields of engineering and vocational education. Our main objective is to provide an assessment of indicators of quality and reliability to aid in decision making. To this end, we recommend an *intuitive* and practical approach, based on mathematical rigor.

The first part of this series (Volume 1) shows the fundamental basis of data analysis in both quality control and in studying the mechanical reliability of materials and structures. Results from laboratory and workshop are discussed in accordance with the technological procedures inherent to the subject matter. We also discuss and interpret the standardization of manufacturing processes as a causal link with geometric and dimensional specifications (GPS, or *Geometrical Product Specification*). This is moreover the educational novelty of this work, which is compared here with consulted praiseworthy publications.

We discuss many laboratory examples, thereby covering a new industrial organization of work. We also use mechanical components from our own real mechanisms, which we built and designed at our production labs. Finite element modification is thus relevant to real machined pieces, controlled and soldered in a dimensional metrology laboratory.

We also discuss mechanical component reliability. Since statistics are common to both this field and quality control, we will simply mention reliability indices in the context of using the structure for which we are performing the calculations.

Scientists from specialized schools and corporations often take interest in the quality of measurement, and thus in the measurement of uncertainties. The so-called cutting-edge endeavors, such as the aeronautics, automotive and nuclear industries, to only mention a few, put an increasing emphasis on the *just measurement*. This text's educational content is noteworthy due to the following:

1) the rigor of the probabilistic methods which support statistical–mathematical treatments of experimental or simulated data;

2) the presentation of varied lab models which come at the end of each chapters: this should help the student to better understand how to:

- define and justify a quality and reliability control target;

- identify the appropriate tools to quantify reliability with respect to capabilities;

- interpret quality (capability) and reliability (reliability indices) indicators;

- choose the adequation test for the distribution (whether justified or used *a priori*);

- identify how trials can be accelerated and their limits;

- analyze the quality and reliability of materials and structures;

- size and *tolerance* (GPS) design structures and materials.

What about uncertainty calculations in applied reliability?

The fracture behavior of structures is often characterized (in linear mechanics) by a local variation of the material's elastic properties. This inevitably leads to sizing calculations which seek to secure the structures derived from the materials. Much work has been, and still is, conducted in a wide range of disciplines from civil engineering to the different variants of mechanics. Here, we do not consider continuum mechanics, but rather probabilistic laws for cracking. Some laws have been systematically repeated to better approach reliability.

Less severe adequation tests would confirm the fissure propagation hypothesis. In fields where safety is a priority, such as medicine (surgery and biomechanics), aviation, and nuclear power plants, to mention but three, theorizing unverifiable concepts would be unacceptable. The relevant reliability calculations must therefore be as rigorous as possible.

Defining safety coefficients would be an important (or even major) element of structure sizing. This definition is costly and does not really offer any real guarantee on safety previsions (unlike security previsions). Today, the interpretation and philosophy of these coefficients is reinforced by increasingly accurate probabilistic calculations. Well-developed computer tools largely contribute to the time and effort of calculation. Thus, we will use software commonly found in various schools (Auto Desk Inventor Pro and ANSYS in modelization and design; MathCAD, GUM, and COSMOS in quality control, metrology, and uncertainty calculations).

Much work has been done to rationalize the concept of applied reliability; however, no "united method" between the mechanical and statistical interpretations of rupture has been defined as yet. Some of the many factors for this non-*consensus* are unpredictable events which randomly create the fault, its propagation, and the ensuing damage. Many researchers have worked on various random probabilistic and deterministic methods. This resulted in many simulation methods, the most common being the Monte Carlo simulation.

In this book, we present some documented applied cases to help teachers succinctly present probabilistic problems (reliability and/or degradation). The intuitive approach takes on an important part in our problem-solving methods, and it is moreover one of the important considerations to present as one contribution through this volume. Many commendable works and books have talked about reliability, quality control, and uncertainty perfectly well, but as separate entities. However, our task here is to verify measurements and ensure that the measurand is well-taught. As Lord Kelvin said, "if you cannot measure it, you cannot improve it". Indeed, measuring identified quantities is an unavoidable part of laboratory life. Theoretical confirmation of physical phenomena must go through *measurement reliability* and its effects on the function are attributed to the material and/or structure, among other things.

Mechanical models (rupture criteria) of continuum mechanics discussed in Chapter 1, Volume 2 make up a reference pool of work used here and there in our case studies, such as the Paris–Erdogan law, the Manson–Coffin law, S-N curves (Wöhler curves), Weibull law (solid mechanics), etc. We could probably (and justly) wonder in what way this chapter is appropriate in works dedicated to reliability. The reason is that these criteria are deliberately targeted. We used them here to avoid the reader having to "digress" into specialized books.

Establishing *confidence* in our results is critical. Measuring a characteristic does not simply mean finding the value of the characteristic. We must also give it an *uncertainty* so as to show the measurement's *quality*. In this book, we will show educational laboratory examples of uncertainty (GUM: Guide to the Expression of Uncertainty in Measurement).

Why then publish another book dedicated to quality control, uncertainties, and reliability?

Firstly, why publish a book which covers two seemingly distinct topics (quality control and reliability including uncertainties)? Because both these fields rely on probabilities, statistics, and a similar method in describing their hypothesis. In quality control, the process is often already known or appears to be under control beforehand, hence the intervention of capability indices (MSP or SPC). Furthermore, the goal is sometimes the competitiveness between manufactured products. Security is shown in secondary terms. Indeed, it is in terms of maintainability and durability that quality control joins reliability as a means to guarantee the functions attributed to a mechanism, component, or even the entire system.

When considering the mechanical reliability of materials and structures, the reliability index is inherently a *safety* indicator. It is often very costly in terms of computation time and very serious in matters of consequence. The common aspect between both fields is still the probabilistic approach. Probabilities and statistical–mathematical tools are necessary to supply theoretical justifications for the computational methods. Again, this book intends to be pragmatic and leaves reasonable room for the intuitive approach of the hypotheses stated throughout.

Finally, we give a brief glossary to standardize the understanding of terms used in dimensional analysis (VIM: Vocabulaire International de Métrologie) and in structural mechanical reliability. This is the best way of having a good agreement on the international terminology of words used to indicate a mesurand, reliability index or even a succinct definition of the capability indicators largely used in quality control.

A. GROUS
November 2012

Chapter 1

Elements of Analysis of Reliability and Quality Control

1.1. Introduction

The past few decades have been marked by a particularly intense evolution of the models of probability theory and of the theory of applied statistics. In terms of the reliability of materials and structures, the fields of application studying the future or replacement of traditional coefficients of security are complicating the empirical and economic sides of the laws of calculation. Regarding the approach to fatigue by fracture mechanics, practice in the laboratory and in construction produces a broad range of problems for which the use of probabilistic and statistical methods proves to be fruitful and sometimes even indispensable. The development of probability theory and of applied statistics, the abundance of new results, and unforeseen catastrophes cause us to pay particular attention to security matters, but with an acute technical and economical sense of concern. From this, comes the *abounding* introduction of reliability analysis methods.

Numerous studies of research have turned toward *probability mechanics*, which aim to predict the behavior of components and to establish a decision-making support system (expert systems). This text commits itself to using the criteria of fracture mechanics to predict, decipher, analyze, and model the reliability of mechanical components. The reliability of structural components uses an ensemble of mathematical and numerical links, which serve to estimate the probability that a component (structure) will reach a certain state of *conventional failure*. This is achieved using the *probabilistic property of resistant elements* of a structure as well as the load that is applied to it.

Contrary to a *prescriptive* classical approach, risk is estimated while maintaining that however *conservative* a regulation may be, it cannot be guaranteed to ensure complete safety. On the other hand, it is necessary for the user of reliability techniques to define what they consider the *failure* of a structure or of a component to entail. If in certain cases that could effectively correspond to the formation of a mechanism of failure, we will define many components and structures as failure criterion, a certain degree of which we will call *damage*.

The important parameters, which include resistance (R) of structure or stresses (S) applied to it, cannot be defined uniquely in terms of nominal or weighted values, but in terms of random variables characterized by their means, their covariance–variances, and their laws of distribution. The estimation of the reliability of a real structure (components) can generally only be approached using an intermediary of a model that is more or less simplified. The analysis of the model will be carried out with the help of mathematical algorithms, which are often given as approximation techniques where rigorous calculation leads to detrimental calculation times.

To evaluate the weak probabilities of component failure, it is normal to begin with the arbitrary integration of probability densities in the domain of failure, i.e. using simulation techniques. The advantage of this way of going about things resides in the necessity of knowing a form explicit to the domain of failure. The disadvantage is the weakness of the speed of convergence. Added to this is the high cost in calculation time. Numerous distribution laws have been used to model the reliability of components and structures. These distributions have not appeared with reliability but with fundamental research relative to diverse fields of application. We will present a few of these. The principal laws of distribution to model the reliability of the components are:

Title	Ordinary by distribution of
Discrete distributions with *finite* support	Bernoulli, discrete uniform, binomial, hypergeometric (Pascal distribution)
Discrete distributions with *countable* support	Geometric, Poisson negative binomial, logarithmic
Continuous distributions with *compact* support	Continuous uniform, triangular, beta
Continuous distributions with *semi-infinite* support	Exponential, gamma (or Erlang), inverse gamma, Chi square (Pearson, χ^2), *Weibull* (2 and 3 parameters), Rayleigh, *log-normal* (Galton), Fisher, Gibbs, Maxwell–Boltzmann, Fermi–Dirac, Bose–Einstein, negative binomial, etc.
Continuous distributions with *infinite support* *Density laws*	*Normal* (Gauss–Laplace), asymmetric normal, Student, uniform, stable, Gumbel student (maxi-mini), Cauchy (Lorentz), Tukey-lambda, *Birnbaum–Saunders* (fatigue), double exponential (Laplace), logistics

Table 1.1. *Main distributions of probability used in reliability and in quality control*

We present below some educational pathways to choose the distributional law correctly and to better model the lifespan of a component or of a structure. Correctly choosing a probabilistic model or experimental data would be in greater harmony with the theoretical justification imposed by distribution law, but is not always a straightforward thing to do. We have, as a test, the exaggeration surrounding Gauss distribution, which is used in almost all theories. Distribution models of lifespan are chosen according to:

– *a physical/statistical argument which corresponds theoretically to a mechanism of failure inherent to a model of life distribution;*

– *a particular model previously used with success for the same thing or a mechanism with a similar fault;*

– *a model which ensures a decent theory/practice applicable to the data relating to the failure.*

Regardless of the method chosen for a significant probabilistic model, it is important to justify the choice fully. For example, it would be inappropriate to approach a model of which the rate of ruin (failure) was constant with an exponential distribution, as this final law is better suited to the sorts of faults that occur when damage is accidental. Galton's (log-normal) and Weibull's models are flexible and fit in well with the fault trends even in cases of weak experimental data. This applies especially when they are projected via the models of acceleration[1] when they are used in very different conditions from test data. These two models are very useful in reducing the failure rates at different magnitudes.

1.1.1. *The importance of true physical acceleration life models (accelerated tests = true acceleration or acceleration)*

Physical acceleration indicates that the functioning of a component produces the same faults, but at a faster rate. Faults can be due to fatigue, corrosion, diffusion, migration, etc. A factor of acceleration is the constant product between two levels of constraints. *True* acceleration occurs when the variation of the constraint is equivalent to the transformation of the timescale in the case of failure. The transformations used are often linear, which implies that "the time of failure" at the high level of constraints will be multiplied by a constant, i.e. by the factor of acceleration, in view of obtaining the time equivalent of failure under the constraint used. We will use the following notations:

1 *When variation of constraints is proportional to the time of failure by a constant, we have true acceleration (i.e. physical acceleration).*

Z_s = time-to-fail at stress

τ_u = time-to-fail at use

$F_s(\tau)$ = cumulative distribution function (CDF) at stress

$F_u(\tau)$ = CDF at use

$f_s(\tau)$ = PDF = (probability density function) at stress

$f_u(\tau)$ = PDF at use

$Z_s(\tau)$ = failure rate at stress

$Z_u(\tau)$ = failure rate at use

1.1.2. *Expression for linear acceleration relationships*

Acceleration factor, F_a, affects constraints according to the following proportions:

$$\left\{\begin{array}{ll} R_u(\tau) = R_s\left(\dfrac{\tau}{F_a}\right) & \rightarrow \text{Reliability} \\ Z_u(\tau) = \left(\dfrac{\tau}{F_a}\right) h_s\left(\dfrac{\tau}{F_a}\right) & \rightarrow \text{Failure Rate} \\ \tau_u = (F_a)\times\tau_s & \rightarrow \text{Time-to-Fail} \\ F_u(\tau) = F_s\left(\dfrac{\tau}{F_a}\right) & \rightarrow \text{Failure Probability} \\ f_u(\tau) = \left(\dfrac{1}{F_a}\right)\times f_s\left(\dfrac{\tau}{F_a}\right) & \rightarrow \text{Density Function} \end{array}\right\} \qquad [1.1]$$

Each failure mode possesses its own *true acceleration*. The failure data must be equally separated by a failure mode according to the pertinence of their inherent true acceleration. If there is acceleration, the data from different units of constraint have the same mode in sampling probability data. *True* acceleration demands the constraint of the physical process causing the change or the degradation that leads to failure. As a rule, different modes of failure will be affected differently by constraints and will have different acceleration factors. It is improbable that a simple factor of acceleration would apply to more than one mechanism of failure. Generally, different modes of failure will be affected by constraints in different ways. They will have different factors of acceleration.

A consequence of the linear acceleration relations shown above is that the form parameter for the key models of life distribution (Weibull and log-normal) does not change for the units functioning under different constraints. The compilations on the probability paper of data of units of different constraints align themselves (form a line) *roughly* in parallel. Some parametric models have successfully been used as population models for time-to-fail resulting in a vast range of failure mechanisms. Sometimes, there are probabilistic *contradictions* based on the physics of failure *modes,* which tend to justify the choice of model.

1.2. Fundamental expression of the calculation of reliability

Introducing reliability as an essential tool in the quality of a component begins at the stage of its very design. It is undeniable that reliability imposes itself as a discipline that rigorously analyzes faults, as it is based on experimental data. As reliability is directly linked to quality, the distributional laws used are often the same or related to each other. We know instinctively that components are numerous and complex in a mechanism (structures). As a result, calculations of reliability become less easily recognizable, as restrictive hypotheses mask them.

The fact that a component is of good *quality* does not indicate that it is *reliable.* There is no correlation between the two, just as *reliability* is not a synonym of *quality*. When it comes to reliability, we make use essentially of the failure rate, of the probability of failure, of security margins, or of reliability indicators. When testing quality, we essentially use machine capabilities [C*m*, C*mk*] or process capabilities [C*p*, C*pk*]. In the following section, we will present the common criteria for reliability, where $F(\tau)$ indicates the function of the probability of failure (or of breakdown) of a component or of a structure assembled in parallel or in series. It represents the probability of having at least one fault before time (τ).

$$F(\tau) = \int_0^{\tau} f(u)\,du \qquad [1.2]$$

where $\mathrm{R}(\tau)$ indicates *reliability* $[0 \leq \mathrm{R}(\tau) \leq 1]$ associated with $F(\tau)$. $\mathrm{R}(\tau)$ is also known as the *survival function* and represents the function probability (service) without fault during period $[0,\ \tau]$. Reliability is defined as the complement of the CDF. Hence, the second term of the following equation:

$$R(\tau) = 1 - F(\tau) = 1 - \int_0^{\tau} f(u)\,du \qquad [1.3]$$

The *failure rate* $\mathrm{Z}(\tau)$ is in fact the relationship between the density of probability of failure in comparison with reliability: $f(\tau)/\mathrm{R}(\tau)$. It is the expression of the probability that a faulty component will fail during $(\Delta\tau)$ under the condition that it does not present a fault before τ. To put it another way, it is the frequency of appearance of failure of a component. The expression of the failure rate is:

$$Z(\tau)=-\frac{f(\tau)}{R(\tau)}=\left(-\frac{1}{R(\tau)}\right)\times\left(\frac{\partial R(\tau)}{\partial\tau}\right)\rightarrow$$
$$\textit{hence } Z(\tau)=-\frac{\partial\ln\left[R(\tau)\right]}{\partial\tau}=\left(-\frac{1}{R(\tau)}\right)\times\left(\frac{\partial R(\tau)}{\partial\tau}\right) \qquad [1.4]$$

$\ln R(\tau)=-\int_0^\tau Z(u)\,du$, so [1.3] will take the form: $R(\tau)=Exp\left[-\int_0^\tau Z(u)\,du\right]$

Average lifetime θ (average time until failure, mean time between failures [MTBF]) is:

$$\theta=\int_0^\infty \tau\times f(\tau)\,d\tau=\int_0^\infty R(\tau)\,d\tau \qquad [1.5]$$

The essential property of *useful life* λ according to an exponential distribution is:

$$\lambda=\left(-\frac{1}{R(\tau)}\right)\times\left(\frac{\partial R(\tau)}{\partial\tau}\right) \quad \textit{i.e.}\quad R(\tau)=Exp^{-\lambda\times\tau} \qquad [1.6]$$

MTB*F* θ is expressed as:

$$\theta=\int_0^\infty \tau\times f(\tau)\,d\tau=\int_0^\infty R(\tau)\,d\tau=\int_0^\infty Exp^{-\lambda\times\tau}\,d\tau=\frac{1}{\lambda} \qquad [1.7]$$

Probability density function $f(\tau)$ represents the probability of failure (fault) of a component at time τ. It is, in fact, the derivative of the CDF $F(\tau)$.

$$f(\tau)=\lambda\times Exp^{-\lambda\times\tau} \qquad [1.8]$$

The *distribution* function (probability of failure, fault) $F(\tau)$ is written as:

$$F(\tau)=1-Exp^{-\lambda\times\tau} \qquad [1.9]$$

Reliability (probability of survival) R(τ) is written as:

$$R(\tau) = Exp^{-\lambda\times\tau} \quad [1.10]$$

The failure rate (fault) is represented as follows:

$$Z(\tau) = \lambda = \text{Constant} = \left(-\frac{1}{R(\tau)}\right)\times\left(\frac{\partial R(\tau)}{\partial\tau}\right) \quad [1.11]$$

Mean	Variance	Standard deviation
$\mu = \theta = 1/\lambda = MTBF$	$V = \sigma^2 = \theta^2 = 1/\lambda^2$	$\sigma = \theta = 1/\lambda$

As an example, exponential distribution allows the calculation of lifespan without fault. What is generally referred to as probability survival is the first term of the Poisson distribution with ($\kappa = 0$).

Estimate: if τ_f is the duration for which an assembly of components will work and κ is the number of faults observed, we will propose the following estimator $\hat{\theta}$:

$$\hat{\theta} = \left(\frac{\tau_f}{\kappa}\right) \; or \; \hat{\lambda} = \left(\frac{1}{\hat{\theta}}\right) = \left(\frac{\kappa}{\tau_f}\right) \quad [1.12]$$

Example of direct application: if $\kappa = 0$, we will consider an estimator of $\hat{\theta}$ the value corresponding to a probability of 0.5, finally obtaining:

$$\hat{\theta}_0 = \left(\frac{\tau_f}{0.69}\right) \; or \; \hat{\lambda} = \left(\frac{1}{\hat{\theta}_0}\right) = \left(\frac{0.69}{\tau_f}\right)$$

Test:

$$Exp^{-\lambda(\tau)} = Exp^{-\lambda\times(M_{bf})} = \frac{1}{2} \Rightarrow M_{bf} = -\frac{\ln(0.5)}{\lambda} \cong \frac{0.69}{\lambda} = 0.69\times\theta$$

Interval of confidence if the trial is *censored* at the threshold $\{1-(\alpha_1+\alpha_2)\}$:

$$\left\{\frac{2\times\tau_f}{\chi^2_{(2v;\alpha_2)}} \prec \theta \leq \frac{2\times\tau_f}{\chi^2_{(2v;1-\alpha_1)}}\right\}; \; \chi^2 = Chi-Square$$

where $\chi^2_{(2r;\alpha_2)}$ *is the* χ^2*test at* $v = 2$ *dof (Degree of freedom)*

Interval of confidence if the trial is *truncated* at the threshold $\{1-(\alpha_1+\alpha_2)\}$:

$$\left\{\frac{2\times\tau_f}{\chi^2_{(2v+1);\alpha_2}} \prec \theta \leq \frac{2\times\tau_f}{\chi^2_{(2v);1-\alpha_1}}\right\};\ \textit{where } \chi^2_{(2v;\alpha_2)} \textit{ is the } \chi^2 \textit{ test at } v=2 \textit{ dof}$$

Numerical application: At 90% confidence level, hence $\alpha_2 = 0.05$ and $1-\alpha_1 = 0.95$, in the censored trial, we will obtain:

$$\left(\frac{2\times 2700}{\chi^2_{(6;\ 0.05)}}\right) = \frac{2\times 2700}{12.6} \prec \theta \leq \left(\frac{2\times 2700}{\chi^2_{(6;\ 0.95_1)}}\right) = \frac{2\times 2700}{1.64} \Rightarrow 428 \prec \theta \leq 3293$$

Using the statistical tables of Karl Pearson's distribution χ^2 (see Appendices, Volume 2 and 3), for the *censored trial*, we will obtain the following:

$$\chi^2_{(4;0.95)} = 0.711 \Rightarrow 365 \prec \theta \leq 6470$$

The most frequently used distribution laws in our case studies are:

– Continuous uniform distribution, U (τ, α, β)

– Discrete uniform distribution U (k, α, β, n)

– Triangular distribution

– Beta distribution B (τ, p, q)

– Normal distribution (Laplace–Gauss)

– Log-normal distribution (Galton)

– Gumbel distribution (Emil Julius)

– Random variable according to one of the Gumbel's distributions

(E_1 *Maximum* and/or E_1 *Minimum*)

– *Weibull* distribution (with two parameters also known as E_2 Min)

– *Weibull* distribution (with three parameters also known as E_3 Min)

– *Birnbaum–Saunders* distribution (*fracture by fatigue*)

– Rayleigh distribution (Lord Rayleigh)

– Rice distribution (*signal treatment*)

– Cauchy distribution (Lorentz)

– Tukey-lambda distribution

– Binomial distribution (Bernoulli's schema)

– Polynomial distribution

– Geometrical distribution

– Hypergeometric distribution (Pascal's)

– Exponential distribution

– Double exponential distribution (Laplace)

– Logistic distribution

– Log-logistic distribution

– Poisson distribution

– Gamma distribution (Erlang)

– Inverse gamma distribution

– Frechet distribution

1.3. Continuous uniform distribution

Continuous uniform distribution is a form of distribution that has continuous density where all intervals with the same length possess the same probability. In terms of its form, it is in fact a generalization of the rectangle function and is often marked as U (α, β). The function of uniform distribution is defined as (α, β) $\in [-\infty, +\infty]$ hence the writing of the support $\alpha \leq \tau \leq \beta$ when the function is written as $f(\tau) = [1/(\beta - \alpha)]$. The following are the main functions of reliability and of quality control that characterize it:

$$\left\{ \begin{array}{l} \textit{Median (center)} = \left(\dfrac{\alpha+\beta}{2}\right) = \textit{Mean} = \mu_T = \left(\dfrac{\alpha+\beta}{2}\right) ; \textit{Mode} = \{a,b\} \\ \textit{Variance} = \sigma_T^2 = \dfrac{(\beta-\alpha)^2}{12} ; \textit{Skewness} = 0 ; \textit{Kurtosis} = -\left(\dfrac{6}{5}\right) \end{array} \right\} \quad [1.13]$$

1.3.1. *Distribution function of probabilities (density of probability)*

$$f_T(\tau) = \begin{cases} \dfrac{1}{\beta - \alpha} \; if \;\; \alpha \leq x \leq \beta \\ 0 \quad\quad if \;\; \tau \prec \alpha, \; or \; \tau \succ \beta \end{cases} \quad [1.14]$$

Graph and calculation of $f(\tau, \alpha, \beta)$

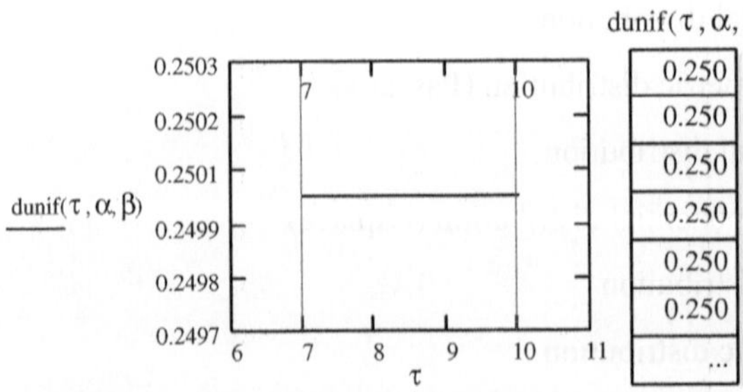

Figure 1.1. *Graph showing the function of the distribution of the continuous U distribution*

1.3.2. *Distribution function*

$$F_T(\tau) = \left\{ \begin{array}{l} \left(\dfrac{\tau - \alpha}{\beta - \alpha} \right) \; if \;\; \alpha \leq x \leq \beta \\ = \begin{cases} 0 & if \;\; x \prec \alpha \\ 1 & if \;\; x \succ \beta \end{cases} \end{array} \right\} \quad [1.15]$$

Graph and calculation showing the distribution function F(τ, α, β):

$$\tau = 7, 7.1, ..., 10 \;\; for \;\; \alpha = 6 \;\; and \;\; \beta = 10 \;\; f(\tau) = \left(\frac{\tau}{\beta - \alpha} \right) sometimes \; f(\tau) = \left(\frac{\tau - \alpha}{\beta - \alpha} \right)$$

Inverse distribution function of cumulative probabilities is given in Figure 1.3:

Figure 1.4. shows *random numbers* between 0 and τ [*rnd* (τ)] which follow a continuous U distribution.

$\text{punif}(\tau, \alpha, \beta) =$

0.250
0.275
0.300
0.325
0.350
0.375
...

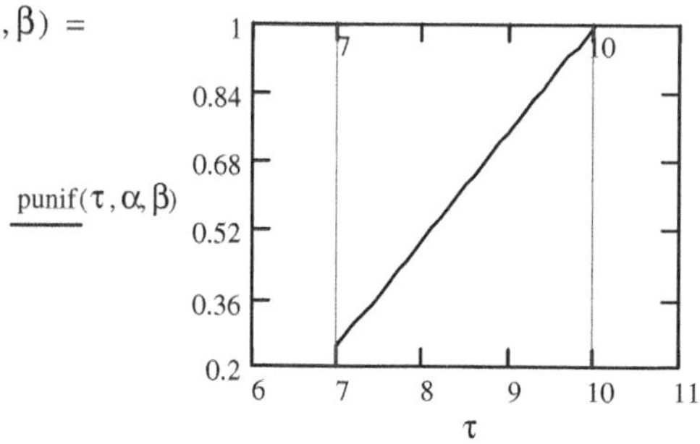

Figure 1.2. *Graph showing the distribution function of the continuous U distribution*

$\text{qunif}(p, \alpha, \beta) =$

6.000
6.800
7.600
8.400
9.200
10.000

$p := 0, 0.2 .. 1$

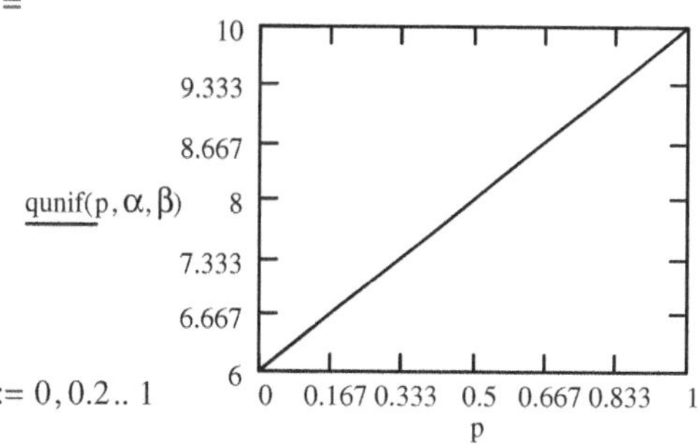

Figure 1.3. *Graph showing the inverse distribution function of continuous U distribution*

$\text{rnd}(\tau) =$

4.74
0.063
1.986
4.292
6.198
3.637
...

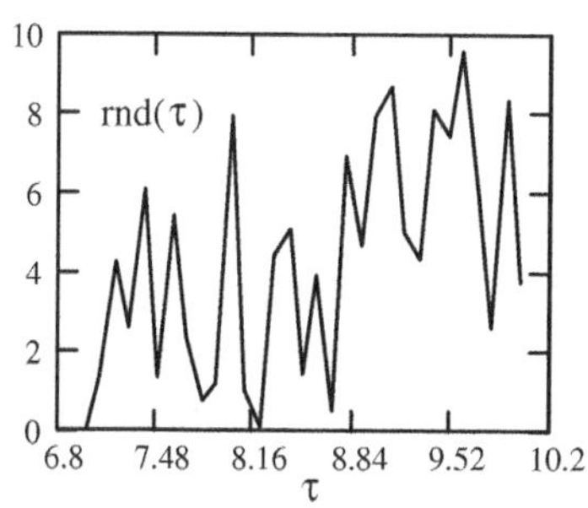

Figure 1.4. *Graph of [rnd (τ)] between 0 and τ [rnd (τ)] following a continuous U distribution*

Random numbers (*m*) which follow a continuous uniform distribution:

$$For\ m = 7 \rightarrow runif(m,\alpha,\beta) = \{7.470;\ 6.875;\ 6.962;\ 9.609;\ 9.611;\ 6.321;\ 6.687\}$$

1.4. Discrete uniform distribution (discrete U)

This is also a probabilistic distribution, which indicates the identical realization of probability in this way (*equiprobability*) to each value of a finite set of possible values. This presents itself according to this schema: A random variable (RV) which can have (n) possible equiprobable values k_1, k_2, k_3,..., k_n (i.e. equal probabilities) follows a uniform distribution. We can see that at any value k_i is equal *to* ($1/n$). It is a classic example of the "throw of the dice" when the score is 1/6. There are cases when the values of this RV which follows discrete U distribution are real. We qualify this in *deterministic* terms by the application of the *distribution function*, formalized below:

Support: $k \in [\alpha, \alpha + 1, \ldots, \beta - 1, \beta]$

$$\left\{\begin{array}{l} Mean = \mu = \left(\dfrac{\alpha+\beta}{2}\right);\ Median = \left(\alpha + \dfrac{n}{2}\right) \\ Skewness = 0\ ;\ Kurtosis = -\left(\dfrac{6(n^2+1)}{5(n^2-1)}\right) \end{array}\right\} \qquad [1.16]$$

$$F(k;\alpha,\beta,n) = \frac{1}{n} \times \sum_{i=1}^{n} H(k-k_i) \qquad [1.17]$$

From the expression $\sum_{i=1}^{n} H(k-k_i)$, we get $H(\tau-\tau_i)$ the distribution function in steps (of Heaviside step). It is, in fact, a determinist distribution centered in τ_0, which is known in (mechanical) physics. It represents the Dirac mass in τ_0. The distribution function $f(\tau, \alpha, \beta)$ (density of probability) of a discrete U distribution on the interval $[\alpha, \beta]$ is written as:

$$f(\tau) = \left\{\begin{array}{ll} \left(\dfrac{1}{\beta-\alpha}\right) & for\ \alpha \le x \le \beta \\ 0 & \text{Otherwise} \end{array}\right\} \qquad [1.18]$$

The distribution function $F(\tau, \alpha, \beta)$ discrete U distribution on the interval $[\alpha, \beta]$ (cumulative probabilities) is then written as:

$$F(\tau)=\left\{\begin{matrix} 0 & for\ \tau<\alpha \\ \dfrac{\tau-\alpha}{\beta-a} & for\ \alpha\leq\tau<\beta \\ 1 & for\ \tau\geq\beta \end{matrix}\right\} \quad [1.19]$$

Parameters of the behavior of discrete U distribution:

$$\left\{\alpha\in\{\ldots-2,\ -1,\ 0,\ 1,\ 2,\ 3\ldots\};\beta\in\{\ldots-2,\ -1,\ 0,\ 1,\ 2,\ 3\ldots\};n=\{\alpha-\beta+1\}\right\}$$

Density of probability of the discrete U function:

$$f(k,\alpha,\beta,n)=\left\{\begin{matrix} \left(\dfrac{1}{n}\right) & for\ \alpha\leq k\leq\beta \\ 0 & \text{otherwise} \end{matrix}\right\} \quad [1.20]$$

Distribution function of the discrete U function:

$$F(k,\alpha,\beta,n)=\left\{\begin{matrix} 0 & for\ k<\alpha \\ \left(\dfrac{k-\alpha+1}{n}\right) & for\ \alpha\leq k<\beta \\ 1 & for\ k\succ\beta \end{matrix}\right\} \quad [1.21]$$

1.5. Triangular distribution

Triangular law can be either maximally or minimally connected to its mode. It has two versions: a discrete distribution and a continuous distribution.

1.5.1. *Discrete triangular distribution version*

The discrete triangular distribution with integer positive parameters α is defined for any integers τ between $-\alpha$ and $+\alpha$ by:

$$P(\tau)=\frac{\alpha+1-|\tau|}{(\alpha+1)^2} \quad [1.22]$$

1.5.2. *Continuous triangular law version*

Continuous triangular distribution on the support [α, β] and of mode γ is defined by the following density on [α, β]. In many fields, triangular distribution is considered as a simplified version of *beta* distribution.

1.5.3. *Links with uniform distribution*

Let X_1 and X_2 be two variables independently and identically distributed according to a uniform law, so that the distribution of the average $Y = \{(X_1 + X_2)/2\}$ is a triangular law of parameters $\alpha = 0$, $\beta = 1$, and $\gamma = ½$. The distribution of the absolute standard deviation $Z = |X_1 + X_2|$ is also distributed according to a triangular distribution of parameters $\alpha = 0$, $\beta = 1$, and $\gamma = 0$.

$$\text{Parameters} \begin{Bmatrix} \alpha \to \alpha \in \{-\infty, +\infty\} \\ \beta \to \beta \succ \alpha \\ \gamma \to \alpha \leq \gamma \leq \beta \end{Bmatrix} \text{ of } support\ \alpha \leq \tau \leq \beta$$

$$Mean\text{: } E(\tau) = \frac{1}{3}(\alpha + \beta + \gamma) \quad Mode = \gamma$$

$$Median\text{: } \begin{Bmatrix} \alpha + \dfrac{\sqrt{(\beta-\alpha)(\gamma-\alpha)}}{\sqrt{2}} & for\ \gamma \geq \left(\dfrac{\beta-\alpha}{2}\right) \\ \beta - \dfrac{\sqrt{(\beta-\alpha)(\beta-\gamma)}}{\sqrt{2}} & for\ \gamma \leq \left(\dfrac{\beta-\alpha}{2}\right) \end{Bmatrix};$$

$$Variance\text{: } E(\tau) = \frac{1}{18}\left(\alpha^2 + \beta^2 + \gamma^2 - \alpha\beta - \alpha\gamma - \beta\gamma\right); \ Kurtosis = -\frac{3}{5}$$

$$Skewness\text{: } E(\tau) = \frac{1}{5}\left(\frac{\sqrt{2}(\alpha+\beta-2\gamma)(2\alpha-\beta-\gamma)(\alpha-2\beta+\gamma)}{\sqrt[3]{\left(\alpha^2+\beta^2+\gamma^2-\alpha\beta-\alpha\gamma-\beta\gamma\right)}}\right);$$

Distribution function (function of mass):

$$f(\tau)=\begin{Bmatrix} \dfrac{2(\tau-\alpha)}{(\beta-\alpha)(\gamma-\alpha)} & \textit{if } \alpha\le\tau\le\gamma \\ \dfrac{2(\beta-\tau)}{(\beta-\alpha)(\beta-\gamma)} & \textit{if } \gamma\le\tau\le\beta \end{Bmatrix} \qquad [1.23]$$

CDF: Formula and graphs:

$$F(\tau)=\begin{Bmatrix} \dfrac{(\tau-\alpha)^2}{(\beta-\alpha)(\gamma-\alpha)} & \textit{if } \alpha\le\tau\le\gamma \\ 1-\dfrac{(\beta-\tau)^2}{(\beta-\alpha)(\beta-\gamma)} & \textit{if } \gamma\le\tau\le\beta \end{Bmatrix} \qquad [1.24]$$

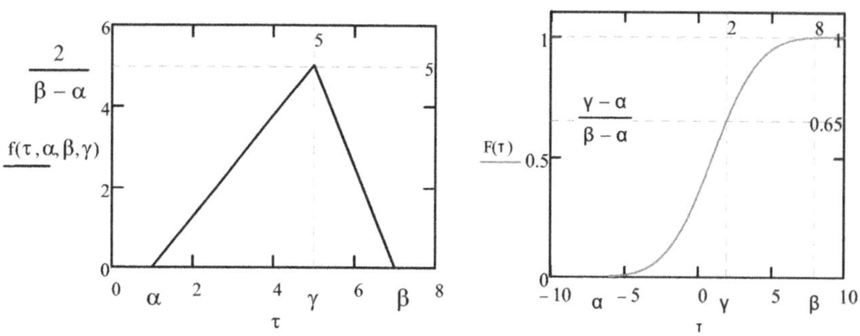

Figure 1.5. *Functions respective of cumulative distribution of triangular distribution*

1.6. Beta distribution

In probability as well as in descriptive statistics, beta distribution comes from a family of continuous probability laws defined in the field as x∈ (0; 1]. With these two distinct parameters (p and q), it is an interesting example of Dirichlet distribution. The main characteristics of the distribution are:

– *Domain* (support): $x\in\{-\infty;+\infty\}$

– $Mean = E(X) = \mu = \left(\dfrac{p}{p+q}\right)$ $Mode = \left(\dfrac{p-1}{p+q-2}\right)$ $with\left(p \; and \; q\right) \succ 1$

– Standard deviation= $\sigma = \sqrt{\dfrac{p \times q}{(p+q)^2 \times (p+q+1)}}$;

– Coefficient of variation $= CV = \sqrt{\dfrac{q}{p \times (p+q+1)}}$

– *Skewness (Asymmetry)* $= A = \dfrac{2(q-p) \times \sqrt{p+q+1}}{(p+q+2) \times \sqrt{p \cdot q}}$

Excess of *kurtosis* of beta distribution is thus written as:

$$Kutosis = 6\left(\frac{p^3 - p^2(2q-1) + q^2(q+1) - 2pq(q+2)}{pq(p+q+2) \times (p+q+3)}\right)$$

1.6.1. *Function of probability density*

The general formula for *the function of probability density* (mass function) of beta distribution is written as:

$$f(\tau) = \frac{(\tau-\alpha)^{p-1} \times (\beta-\tau)^{q-1}}{B(p,q) \times (\beta-\alpha)^{p+q+1}} \quad with\ \alpha \le \tau \le \beta\ ;\ (p,\ q\ ) \succ 0 \qquad [1.25]$$

The function of distribution from the MathCAD program dbeta (τ, p, q) gives results for treating random variables (τ) directly.

$$f(\tau) = beta(\tau, p, q) = \left(\frac{\Gamma(p+q)}{\Gamma(q) \times \Gamma(q)}\right) \times \left(\tau^{p-1}\right) \times (1-\tau)^{q-1};\ (p, q\ ) \succ 0 \qquad [1.26]$$

In this distribution, p and q are the (real) shape parameters. In the case where p = 0 and q = 1, the distribution is a *beta standard* distribution which is written as:

$$f(\tau) = \frac{(\tau)^{p-1} \times (1-\tau)^{q-1}}{B(p,q)} \quad with\ 0 \le \tau \le 1\ ;\ (p,\ q\ ) \succ 0 \qquad [1.27]$$

The density of beta distribution can take different forms according to p and q:

For p and q: … ↓	The distribution is: …. ↓
$p < 1$ and $q < 1$→	In a U shape
$p < 1$ and $q \geq 1$ or $P = 1$ and $b > 1$→	Is strictly falling
$p = 1$ and $q > 2$ →	Is strictly convex
$p > 2$ and $q = 1$→	Is strictly convex
$p = 1$ and $q = 2$ →	Is a straight line
$p = 2$ and $q = 1$ →	Is a straight line
$p = 1$ and $1 < q < 2$ →	Is strictly concave
$1 < p < 2$ and $q = 1$ →	Is strictly concave
$p = 1$ and $q = 1$ →	Is a continuous uniform distribution
$p = 1$ and $q < 1$ or $p > 1$ and $q \rightarrow \leq 1$	Is strictly rising
$p > 1$ and $q > 1$→	Is unimodal
$p = q$ → density around ½	Is symmetrical

Table 1.2. *Density of beta distribution in different forms according to p and q*

For: τ from (0 1] and for the distinct parameters (p and q) of the beta distribution, we propose:

$p_1 = ½$ and $q_1 = ½$	$p_2 = 3.5$ and $q_2 = 1.5$	$p_3 = 1.5$ and $q_3 = 3.5$
$p_4 = 3.5$ and $q_4 = 3.5$	$p_5 = 2.5$ and $q_5 = 5.5$	$p_6 = 7.5$ and $q_6 = 3.5$

The resulting graph from this allows us to read (and to see) the following pattern:

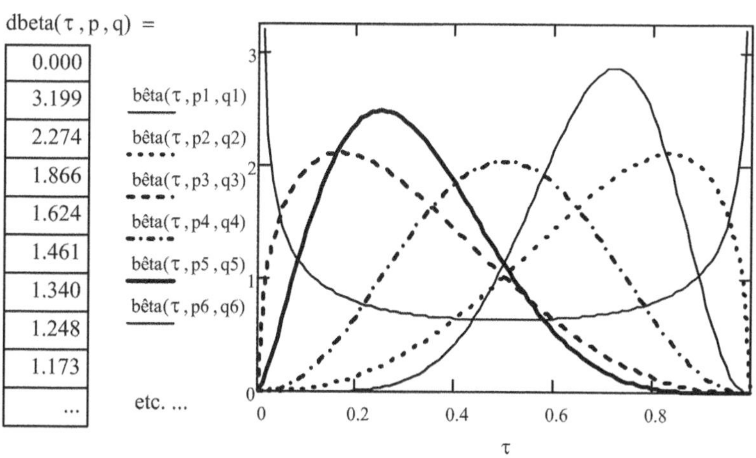

Figure 1.6. *Graph showing the beta distribution according to shape parameters (p, q)*

1.6.2. *Distribution function of cumulative probability*

$$F(\tau)=I_\tau\left(p,q\right)=\frac{\int_0^\tau t^{p-1}\left(1-t\right)^{q-1}dt}{B\left(p,q\right)} \quad \textit{with } 0\le\tau\le 1\,;\ \left(p,\ q\ \right)\succ 0 \qquad [1.28]$$

where $B(p, q)$ is the beta function.

$$F(\tau,p,q)=I_\tau\left(p,q\right)=\frac{B_\tau\left(p,q\right)}{B\left(p,q\right)} \quad \textit{with } 0\le\tau\le 1\,;\ \left(p,\ q\ \right)\succ 0 \qquad [1.29]$$

where B_τ (p, q) is the incomplete beta function and I_τ (p, q) is the incomplete regularized beta function. The pbeta *CDF pbeta* (τ, p, q) directly provides the results of the treatment of the random variable (τ).

Beta distribution is not used in mechanical reliability. We deliberately have not followed conventional formulae. The reader will be able to refer to specialized manuals in this field for more detailed information.

pbeta(τ, p1, q1) =

0.000
0.064
0.090
0.111
0.128
0.144
0.158
0.170
0.183
...

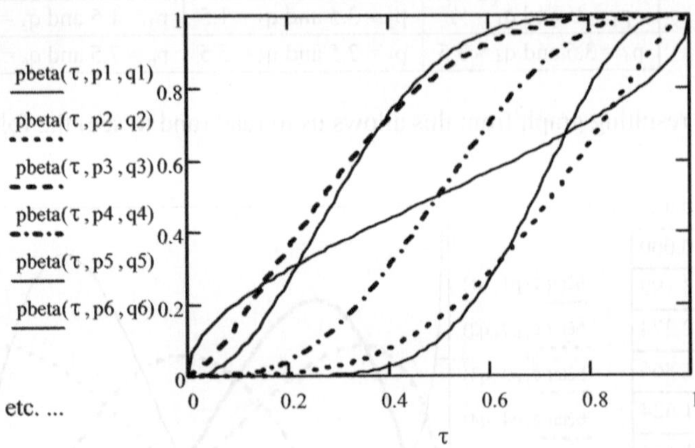

Figure 1.7. *Graph showing cumulative beta distribution according to shape parameters (p, q)*

Density of inverse probability of *beta* distribution *qbeta* (t, p, q). For $t = 0$ to 1, we will have:

pbeta(τ,p1,q1) =

0.000
0.064
0.090
0.111
0.128
0.144
0.158
0.170
0.183
...

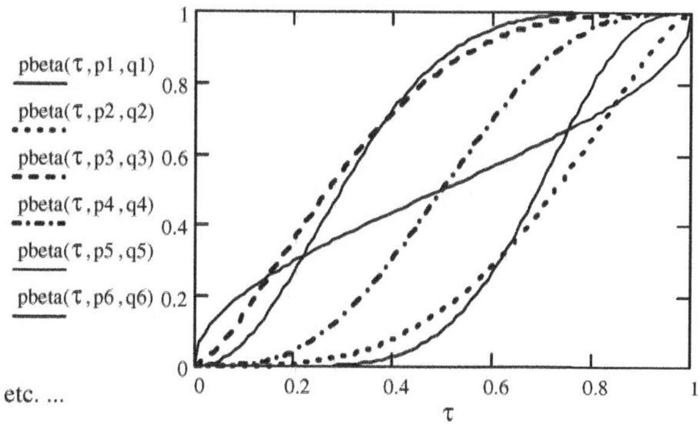

Figure 1.8. *Graph showing the beta inverse cumulative distribution according to shape parameters (p, q)*

The vector of (***m***) random numbers having beta distribution probability is written as **r**beta (**m**, p, q). (***r***) for random numbers, where ***m*** is a positive integer.

For $m = 7$, we have the vectors according to the parameters of beta distribution (p, q):

$$rbeta\left(m, p_1, q_1\right) = \{\ 0.148,\ 0.287,\ 0.975,\ 0.213,\ 0.361,\ 0.583,\ 0.282\ \}$$

$$rbeta\left(m, p_2, q_2\right) = \{\ 0.955,\ 0.715,\ 0.422,\ 0.819,\ 0.483,\ 0.766,\ 0.723\ \}$$

$$rbeta\left(m, p_3, q_3\right) = \{\ 0.054,\ 0.407,\ 0.070,\ 0.087,\ 0.434,\ 0.210,\ 0.150\ \}$$

1.6.3. *Estimation of the parameters (p, q) of the beta distribution*

According to empirical law, we propose:

$$\left\{ \mu = \bar{x} = \left(\frac{1}{n}\right)\sum_{i=1}^{n} x_i \ \ and\ \ VAR = \left(\frac{1}{n}\right)\sum_{i=1}^{n}\left(x_i - \bar{x}\right)^2 \right\}$$

Variance: The method of moments gives the following estimations:

$$p = \bar{x}\left(\frac{\bar{x}\left(1-\bar{x}\right)}{VAR} - 1\right) \text{ and } q = \left(1-\bar{x}\right)\left(\frac{\bar{x}\left(1-\bar{x}\right)}{VAR} - 1\right)$$

1.6.4. *Distribution associated with beta distribution*

– If Z has beta distribution, the random variable $Z = X/(1-X)$ is distributed according to beta distribution of the second order (or type). Beta distribution (1.1) is identical to continuous uniform distribution;

– If, for example X and Y are independently distributed according to gamma distribution, with parameters (p, θ) and (q, θ), respectively, then the random variable $X/(X+Y)$ is distributed according to beta law (p, q);

– If $X \approx U(0;1)$ according to a uniform distribution: $X_2 \approx$ beta(1/2; 1) then the *k*th statistic of the order of an *n* sample of uniform U distributions (0.1] follows beta distribution $\{k, n-k+1\}$.

1.7. Normal distribution

Normal distribution is the most well-known type of distribution. Ever since Moivre (1738), this distribution has been used in reliability and in quality control as a limit of binomial distribution. We also use it as a model for the distribution of *measurement errors*, which in metrology is conventionally called a "true" value. It also plays an important role in the asymptotic behavior of other probability distributions.

In reliability, Gauss' distribution is widely used to represent the distribution of the lifespan of components toward the end of their useful life (*component fatigue*). We can find the explanation for this from looking at ever-increasing failure rates, Z(τ). The literature suggests only using this in reliability if average lifespan is superior to at least three times the standard deviation.

1.7.1. *Arithmetic mean*

For a grouping of (n) values by the class of distribution frequencies, we propose the following as the arithmetic mean:

$$\left\{ \mu = \sum_{i=1}^{n} n_i \times x_i \Big/ n \quad or \quad \mu = \sum_{i=1}^{n} x_i \Big/ n \right\}$$

with (n_i) the size of central class x_i. In a situation where the (n) values (x_i) are not grouped by class of statistical series, we will use the expression of variance. We will do the same for the expressions of variance. For a grouping of (n) values by class of distribution frequencies, we propose the following arithmetic mean:

$$V=\sigma^2=\sum_{i=1}^{n} n_i\times\left(x_i-\mu\right)^2 \Big/ \sum_{i=1}^{n} n_i$$

By analogy of the above, in a situation where the (n) values (x_i) are not grouped by class according to statistical series, we will use this expression, with N, as a unit of full size or balance of (V):

$$V=\sigma^2=\frac{\sum_{i=1}^{n}\left(x_i-\mu\right)^2}{N}\rightarrow\sigma=\sqrt{\frac{\sum_{i=1}^{n}\left(x_i-\mu\right)^2}{n-1}}$$

For the estimation of sample data, (n) represents the extent of sample data that is not grouped by class. The calculations, which present this, are shown below. In cases where data sample sizes do not exceed 15 values, we recommend scaling up the calculations based on Table 1.3 (from the literature):

n	β_m	n	β_m	n	β_m	n	β_m
2	1.254	**6**	1.050	**10**	1.028	**14**	1.020
3	1.128	**4**	1.043	**11**	1.025	**15**	1.018
4	1.085	**8**	1.036	**12**	1.023		
5	1.063	**9**	1.031	**13**	1.021		

Table 1.3. *Table of values of β_m (see Volume 2 and 3 Appendices Table A.1 and formula)*

For a sample size **<15** values, we will calculate the following: $\sigma^*=\sigma\times\beta_m$. The normal centered variable is reduced to:

$$u=\left(\frac{\tau-u}{\sigma}\right)\rightarrow\varphi(u)=\frac{1}{\sqrt{2\pi}}\times Exp\left\{-\frac{u^2}{2}\right\}\ with\ (\mu,\sigma)=[0,1] \qquad [1.30]$$

In fact, it corresponds to Gauss' distribution with a mean = 0 and a unitary standard deviation σ = 1. The tables from statistics (see Volume 2 Appendices Tables A.5 and A.6) allow us to read the tabulated values of this law. Its CDF will be written as:

$$\Phi(u)=\int_0^u\varphi(x)dx \qquad [1.31]$$

Probability density (distribution function of probability) is:

$$f_X(x) = \frac{1}{\sigma_X \times \sqrt{2\pi}} \times Exp\left\{-\frac{(x_i - \mu_X)^2}{2 \times \sigma_X^2}\right\} \quad [1.32]$$

1.7.2. *Reliability*

As we know that the normal variable is defined between [–∞ and +∞], the use of this distribution seems completely restrictive, as it has been proven to be impossible to postulate a negative lifespan.

$$R(\tau) = 1 - \Phi\left(\frac{\tau - \mu}{\sigma}\right) = \Phi\left(\frac{\mu - \tau}{\sigma}\right) \quad [1.33]$$

Function of probability density for reduced centered distribution is:

$$f(\tau) = \left(\frac{1}{\sigma}\right) \times \varphi\left(\frac{\tau - \mu}{\sigma}\right) \quad [1.34]$$

CDF (cumulative probability) takes form [1.35], which represents *failure rate*:

$$Z(\tau) = \left(\frac{1}{\sigma}\right) \times \varphi\left(\frac{\tau - \mu}{\sigma}\right) \Big/ \varphi\left(\frac{\mu - \tau}{\sigma}\right) \quad [1.35]$$

Due to symmetry, we can note the following: $F_x(x) = 1 - F_x(x)$. For example, $F_x(3) = P\{-3 \leq x \leq 3\} = 0.997$, i.e. nearly the entire probability is included between [–3 and +3].

The curves representing these functions are shown on the following graphs:

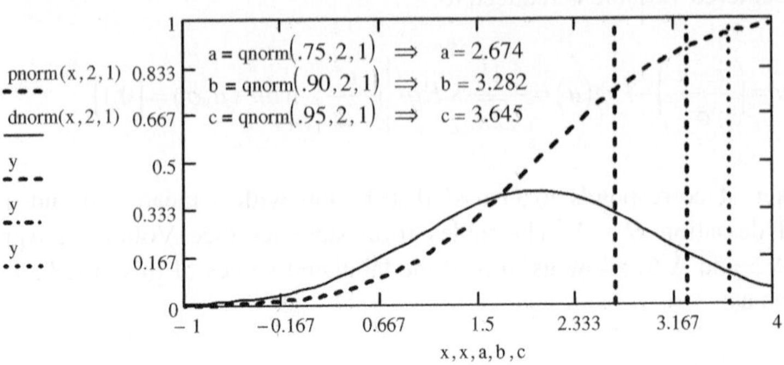

Figure 1.9. *Graphs showing cumulative distribution functions*

1.7.3. *Stabilization and normalization of variance error*

Looking at the multivariable data of variance s^2, variance *error* (ε) increases with one of the independent variables, *P.* Lorczak (see MathCAD 15) already resolved this problem. This inspired us to resolve our own mechanics problem with the error done on the variance for n = 56 points using a mechanism that originated in mechanical design. The calculation of the standard deviation σ_{INCR} is programmed below using the MathCAD program.

n = 56

sample size: n

```
σINCR := | DATA⟨0⟩ ← runif(n,0,1)
         | DATA⟨1⟩ ← runif(n,0,1)
         | "Two columns of random data for x1 and x2"
         | σincr ← DATA⟨0⟩
         | "create a column of errors that depend on the first data column"
         | ε ← vectorize(σincr·rnorm(n,0,1))
         | β ← (4, 7, 2)ᵀ
         | "create a linear relationship between x1, x2, and a dep. variable"
         | DATA⟨2⟩ ← β0 + β1·DATA⟨0⟩ + β2·DATA⟨1⟩ + ε
         | DATA
```

The standard deviation function $y = f(\sigma_{INCR})$ is written as:

$$y = \sigma INCR^{\langle 2\rangle} \;;\; i = 0 ... last(y) \;;\; X^{\langle 0\rangle} = \sigma INCR^{\langle 0\rangle} \;;\; X^{\langle 1\rangle} = \sigma INCR^{\langle 1\rangle}$$

The term (ε), for error, comes from a normal distribution with no mean (μ = 0). To verify the hypothesis of the model, we will bring in a polynomial multivariable with two columns of independent variables and with a singular column relative to the dependent variables:

$$params = regress\{X, y, 2\} \;;\; yfits_i = \text{int}\,erp\left\lfloor params, X, y, \left(X^T\right)^{\langle i\rangle} \right\rfloor \qquad [1.36]$$

$resids = \{y - yfits\}$; the expression of error is: $stERR = \sqrt{\dfrac{\sum resids^2}{last(y) - 5}}$ [1.37]

The numerical value of the error is the following: $stERR = 0.535089$ (54%).

If comparing standard residual dispersion and each of the independent variables or the predicted values does not bring any perceptible model to light, the variance of the error terms is likely to be constant. As for comparing residual dispersions with x_1 (Figure 1.10), the points disperse increasingly from left to right, indicating that error (ε) increases with x_1.

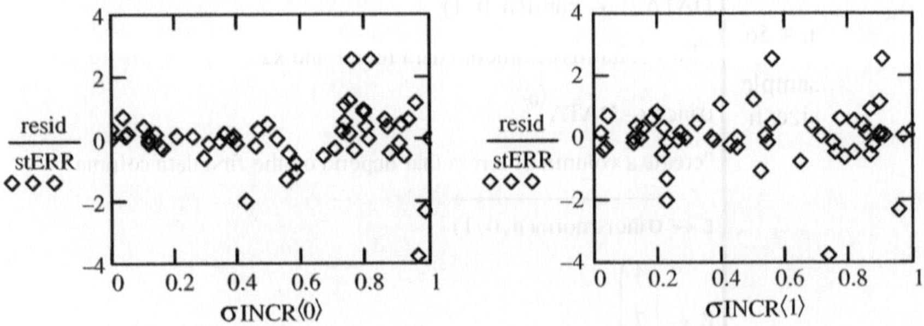

Figure 1.10. *Distribution of errors (on variance) in normal distribution*

On a similar graph for x_2, we clearly see (looking at the points in a lozenge shape in the above graph) that the model is not *clear*. Elsewhere, the model is once again evident on the graph of residual dispersion (circular points on the graph) for the predicted values:

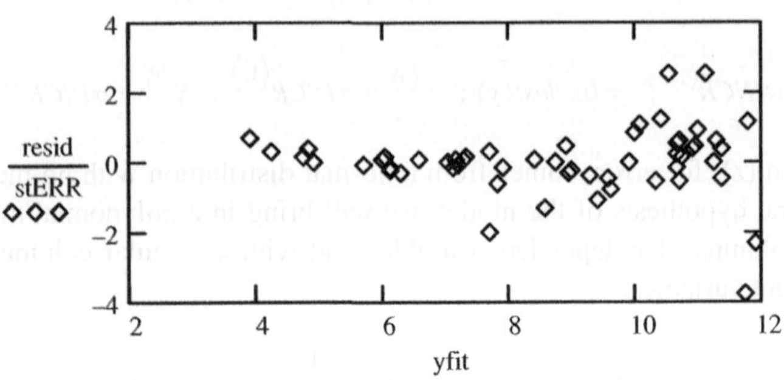

Figure 1.11. *Distribution of errors in normal distribution*

To obtain a valid regression, we must neutralize the increasing difference in error while transforming the dependent data. The difference stabilizing the transformation is represented above in order of rising severity.

$$\left\{ y_{sta} = \overrightarrow{\left(\sqrt{y}\right)} \; ; \; y_{sta} = \overrightarrow{\left(\ln y\right)} \; ; \; y_{sta} = \overrightarrow{\left(\frac{1}{y}\right)} \right\} \qquad [1.38]$$

Above, we have determined new residues against the predicted values, to show the effect of stabilization more clearly (see the graph on which the points are represented by little blue squares).

$$params = regress\left\{X, y_{sta}, 2\right\} \; ; \; yfits_i = \text{int}\,erp\left[params, X, y_{sta}, \left(X^T\right)^{\langle i \rangle} \right]$$

$$resids = \left(y_{sta} \times \text{yfits}_i\right) \text{ the expression of the error is stERR} = \sqrt{\frac{\sum resids^2}{last(y) - 5}}$$

The variance error will be = 0.535089, which is shown in the following graph:

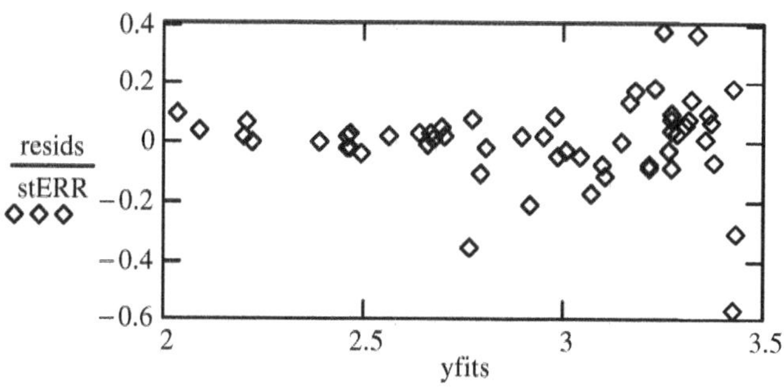

Figure 1.12. *Calculation of residues compared to variance error*

It is noticeable that certain *transformations* are quite strong. They actually accentuate error. For example, the graph *has the task of presenting a curve* rather than presenting purely random values, thus indicating a non-linear model. This last model could provide a fairer cumulative distribution of data (experimental in our area of study). It is important to correct distributional abnormalities (primarily Gaussian) to a certain extent. Stabilizing the wide dispersion of error and verifying "*non-normality*", we will now consider a population sample with well-known terms of error which are exponentially distributed:

$$\varepsilon\text{EXPO} := \left| \begin{array}{l} \text{DATA}^{\langle 0\rangle} \leftarrow \text{runif}(\text{n\_expo}, 0, 1) \\ \text{DATA}^{\langle 1\rangle} \leftarrow \text{runif}(\text{n\_expo}, 0, 1) \\ \varepsilon 1 \leftarrow \text{rexp}(\text{n\_expo}, 10) \\ \beta \leftarrow \begin{pmatrix} 4 \\ 7 \\ 2 \end{pmatrix} \\ \text{DATA}^{\langle 2\rangle} \leftarrow \beta_0 + \beta_1 \cdot \text{DATA}^{\langle 0\rangle} + \beta_2 \cdot \text{DATA}^{\langle 1\rangle} + \varepsilon 1 \\ \text{DATA} \end{array} \right.$$

n, sample size:

n_expo = 25

$$ye = \varepsilon EXPO^{\langle 2\rangle};\ i = 0...last(ye);\ Xe_i^{\langle 1\rangle} = 1\ Xe^{\langle 1\rangle} = \varepsilon EXPO^{\langle 0\rangle}\ Xe^{\langle 2\rangle} = \varepsilon EXPO^{\langle 1\rangle}$$

In Chapter 2, we will present a *t-test* of mean-normality. We will deal with the localized estimator of the mean of normal distribution in Chapter 3 where more highly detailed cases will be closely examined. Non-normality may be detected on an ordinary curve of normalized (standardized) residues. If the graph resembles a straight line, we must conclude that these errors are probably due to a normal distribution. We recognize this fact in quality control when looking at Henry's constant (see Chapters 2, 8 of Volume 2, and 1 of Volume 3).

$$params = \left(X_e^T \times Xe\right)^{-1} \times \left(X_e^T \times ye\right) \textit{ and } \text{yfit}{=}Xe \times params \textit{ then } params = \begin{pmatrix} 3.993042 \\ 7.015407 \\ 2.187455 \end{pmatrix}$$

$$reside = \left(ye \times \text{yfit}\right) \textit{ and } \text{stERR}{=}\sqrt{\frac{\sum reside^2}{last\left(ye\right) - cols\left(X_e\right)}} \textit{ then } \text{stERR} = 0.127297$$

As expected, the residues are not normally distributed, as illustrated on the graph of our non-linear results from the calculations. Once again, we have defined the transformations in order of rising severity. To see the effect of other transformations, we have proceeded using the following mathematical method:

$$\left\{ y_{norm} = \sqrt{ye}\ ;\ y_{norm} = \ln\left(ye\right)\ ;\ y_{norm} = \frac{1}{ye} \right\} \qquad [1.39]$$

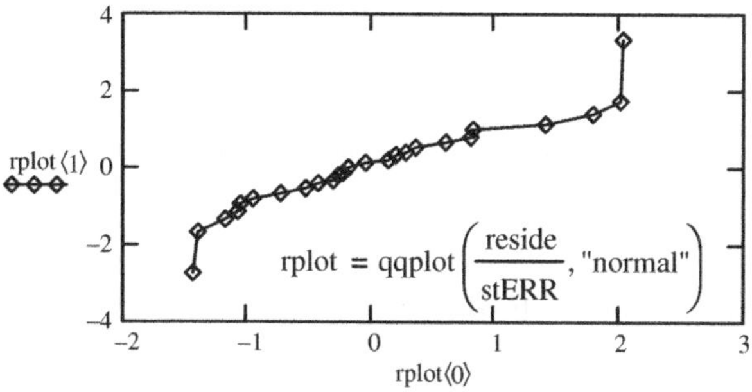

Figure 1.13. *Graph showing the line of non-normality (proof)*

The next graph-checks, used to either confirm or deny the residual (standardized) normality, yielded:

$$params = \left(X_e^T \times Xe\right)^{-1} \times \left(X_e^T \times norm\right) \text{ and yfit=} Xe \times params \text{ then } params = \begin{pmatrix} 2.133535 \\ 1.192698 \\ 0.368536 \end{pmatrix}$$

This allowed us to obtain the results of the residual error as:

$$reside = \left(y_{norm} \times \text{yfit}\right) \text{ and stERR=}\sqrt{\frac{\sum reside^2}{last\left(y_{norm}\right) - cols\left(X_e\right)}} \text{ then stERR} = 0.028731$$

Figure 1.14 shows the graph connected to Henry's constant:

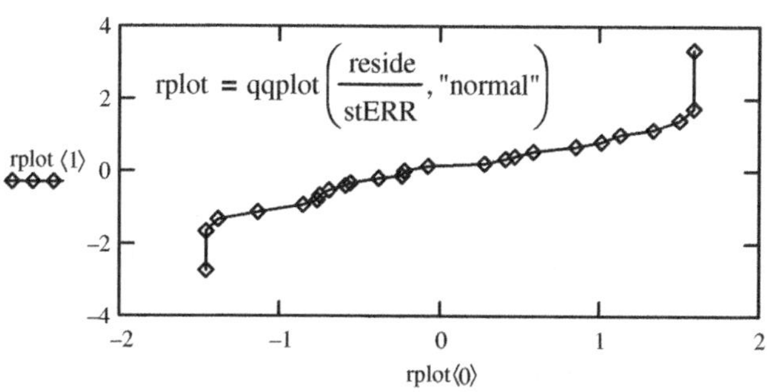

Figure 1.14. *Graph showing the line of non-normality (test f)*

We can examine the relationship between transformation and non-normality in terms of error by taking into account other equations. Once again, we notice that certain transformations may work better than others. In this final graph, we move even closer to getting a straight line, which would claim or declare normality to be null.

1.8. Log-normal distribution (Galton)

This distribution is characterized by its use of the estimation of the component lifespan, if the logarithm (or log) of longevity of the component follows a normal distribution. However, the domain is never negative. For this reason, there is no limit on the use of this distribution for reliability. Failure rate (risk) increases from the beginning of a components life (structure) and it has quite a dissymmetrical distribution. This distribution is widely used in material and structure reliability and in repairs, to list only a few of its uses:

– *Mean*:

$$\mu_{(\tau)} = Exp\left(\mu + \frac{\sigma^2}{2}\right) \text{ or inversely } \mu = 2\ln\mu_{(\tau)} - \frac{1}{2\ln\left(\mu_{(\tau)}^2 + \sigma_{(\tau)}^2\right)}$$

– *Variance*:

$$\sigma_{(\tau)}^2 = \exp\left(2\mu + 2\sigma^2\right) - Exp\left(2\mu + \sigma^2\right) \text{ or inversely } \sigma^2 = \ln\left(\mu_{(\tau)}^2 + \sigma_{(\tau)}^2\right) - 2\ln\mu_{(\tau)}$$

– *Probability density (for the Neperian logarithm)*:

$$f(\tau) = \frac{1}{\tau \cdot \sigma \cdot \sqrt{2\pi}} Exp^{\left(-\frac{1}{2}\left(\frac{\ln\tau - \mu}{\sigma}\right)^2\right)} \quad \{\mu \text{ of } y = \ln\tau \text{ and } \sigma \text{ of } y = \ln\tau\} \quad [1.40]$$

– *CDF*:

$$F(\tau) = \Phi\left(\frac{\ln\tau - u}{\sigma}\right) \quad [1.41]$$

where Φ (u) is the *CDF* of the random centered reduced variable (tabulated).

1.9. The Gumbel distribution

The Gumbel distribution is a distribution of *continuous* probability used in scientific and technical activities such as hydrology. It is a distribution that is said to have extreme values and it is also used to model critical events like natural disasters such as earthquakes. We must remain realistic, however, as this law does not predict anything physical. It allows the modeling of probabilistic behavior while using data collected by experience and real life.

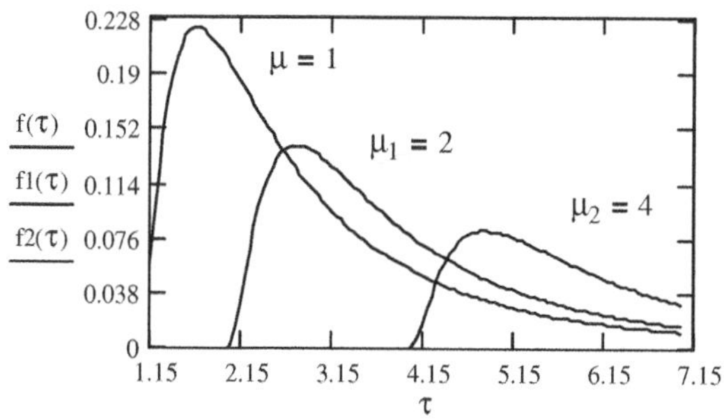

Figure 1.15. *Graph showing distribution functions of log-normal distribution*

1.9.1. *Random variable according to the Gumbel distribution (CRV, E_1 Maximum)*

The mean and the standard deviation take the following forms, respectively:

$$\mu(x) = (\alpha \times \varepsilon) + \beta \text{ , } with\ \varepsilon = \text{ constant } and\ \sigma(x) = \left(\frac{\pi \times \alpha}{\sqrt{6}}\right)$$

where:

– X ∈ R, hence [- ∞ < x < + ∞];

– ε is a constant called Euler-Mascheroni = 0.5772156649;

– α is a parameter in a real position;

– β > 0 is a scale parameter.

The form of the *probability density* defined in **R** is the following:

$$f(x) = \frac{1}{\alpha} \times \left\{ Exp^{\left(-\frac{x-\beta}{\alpha}\right) - Exp^{\left(-\frac{x-\beta}{\alpha}\right)}} \right\} \quad when \quad \alpha \succ 0 \qquad [1.42]$$

Its CDF is written as:

$$F(x;\mu,\beta)=F(x)=\exp^{Exp^{\left(-\frac{x-\beta}{\alpha}\right)}} \quad with \quad \alpha \succ 0 \qquad [1.43]$$

f(x) =
9.2428·10^{-3}
9.3889·10^{-3}
9.5371·10^{-3}
9.6874·10^{-3}
9.8400·10^{-3}
9.9948·10^{-3}
0.0102
0.0103
...

When α = 6 β = 3 x := −14, −13.9 .. 20

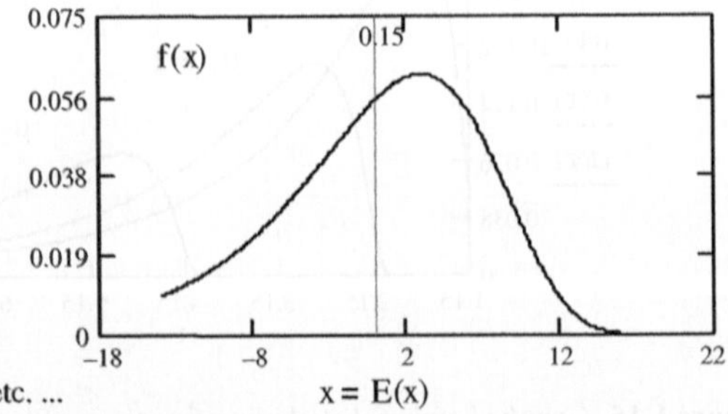

etc. ...

ε = 0.5772156649

Figure 1.16. *Graph showing the function of distribution of E_1 Max*

1.9.2. *Random variable according to the Gumbel distribution (CRV E_1 Minimum)*

The mean and the standard deviation take the following forms, respectively:

$$\mu(x)=\beta-(\alpha\times\varepsilon) \quad with \ \varepsilon = \text{ constant, } and \ \sigma(x)=\left(\frac{\pi\times\alpha}{\sqrt{6}}\right)$$

If random independent variables follow a Gaussian center, their maxima follow. In *n*'s case, this is quite high, Gumbel's distribution of parameters with position *α*, a reality > 0) and scale β.

$$\alpha=\sqrt{2\times\ln\times(n)} \ \text{ and } \ \beta=\frac{1}{\sqrt{2\times\ln\times(n)}} \quad (i.e., \ n \ years) \qquad [1.44]$$

The form of probability *density* defined in **R** is shown in the following:

$$f(x)=\frac{1}{\alpha}\times\left\{Exp^{\left(\frac{x-\beta}{\alpha}\right)-Exp^{\left(\frac{x-\beta}{\alpha}\right)}}\right\} \quad When \ \alpha \succ 0 \qquad [1.45]$$

f(x) =

$9.2428 \cdot 10^{-3}$
$9.3889 \cdot 10^{-3}$
$9.5371 \cdot 10^{-3}$
$9.6874 \cdot 10^{-3}$
$9.8400 \cdot 10^{-3}$
$9.9948 \cdot 10^{-3}$
0.0102
0.0103
...

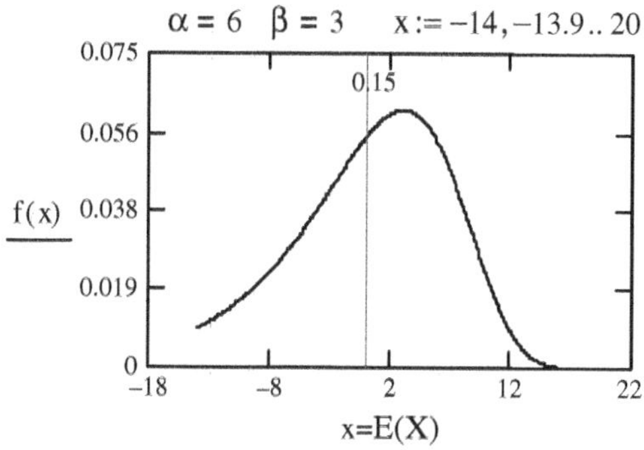

ε = 0.5772156649

Figure 1.17. *Graph showing the distribution function E_1 Min*

Its CDF is written as:

$$F(x;\mu,\beta) = F(x) = 1 - E\,\mathrm{xp} - E\,\mathrm{xp}\left\{\left(\frac{x-\beta}{\alpha}\right)\right\} \textit{with} \quad \alpha \succ 0 \qquad [1.46]$$

With N(μ; β) and where μ = 0 and β = 1, we obtain a *standard* Gumbel distribution.

1.10. The Frechet distribution (E_2 Max)

This law is also called E_2 max. It is characterized by:

– *Mean*:

$\mu(x) = \left(\frac{1}{\beta}\right) \times \Gamma\left(1 - \frac{1}{\alpha}\right)$ with $\Gamma(.)$ a Eulerian function of the second type

$$\Gamma(x) = \int_0^{+\infty} \tau^{x-1} \times \exp^{-\tau} d\tau \; ; \; x \succ 0$$

– *Standard deviation*:

$$\sigma(x) = \sqrt{\left(\frac{1}{\beta^2}\right) \times \left\{\Gamma\left(1 - \frac{2}{\alpha}\right) - \Gamma^2\left(1 - \frac{1}{\alpha}\right)\right\}} ; x \succ 0$$

– *Probability density of variables (x)*:

$$f(x) = \alpha \times \beta \times (x \cdot \beta)^{-(\alpha+1)} \times E\,\mathrm{xp}^{-(\beta \cdot x)^{-\alpha}} \quad \textit{when } \alpha \succ 0 \textit{ and } \beta \succ 0 \qquad [1.47]$$

– *CDF of variables (x)*:

$$F(x) = E\,\mathrm{xp}^{-(\beta \cdot x)^{-\alpha}} \quad \textit{with } \alpha \succ 0 \textit{ and } \beta \succ 0 \qquad [1.48]$$

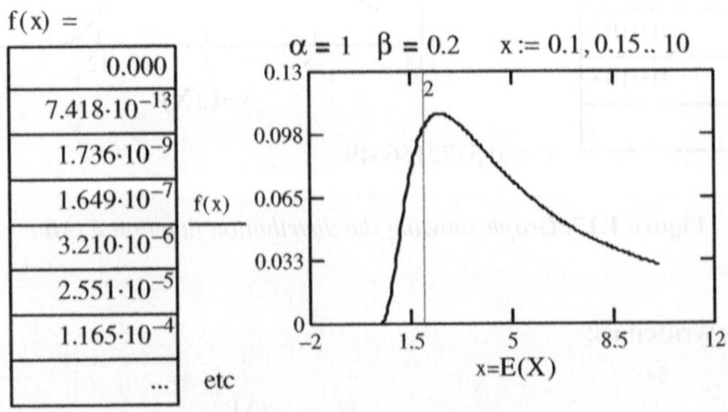

Figure 1.18. *Graph showing Frechet's E_2 Max distribution function*

1.11. The Weibull distribution (with three parameters)

This law has been taken up in many scientific fields and is regularly used in reliability and quality control (see Appendix to Volume 2 of this series, Figures A.1 and A.2) to describe the lifespan of a component or structure, as well as to describe the time of the first failure (MTBF *between and before failure*) or beyond the passage of time after consecutive failures (*after failure*). In mechanics, this law has the advantage of taking into account aging phenomena and is advantageous compared to other laws because it provides a better characterization of systems solicited by mechanical degradations across the three stages of life. The Galton distribution, for example, has a disadvantage in this respect, as it presents a failure rate that is initially at zero and then increases, heading for its maximum, to then move toward zero for high values of τ.

The Weibull distribution also covers exponential distribution for $\beta \approx 3.6$. It is sometimes for this reason that we call the Weibull distribution a *chameleon*, due to its adaptive flexibility. In cases of fissuring due to fatigue, we recommend the use of the *Birnbaum–Saunders* distribution even though it is specifically restrained. For stochastic processes of destruction by fissuring, it is possible to calculate the distribution of fracture-time. The Weibull distribution initially introduced by

J. Frechet and then again by W. Weibull is widely used in mechanics because of its form, including normal distribution, exponential distribution, and log-normal distribution, according to the value of the parameter in the form (β) which makes it up. Moreover, it is a flexible law that depends on three parameters: $\gamma > 0$, $\beta > 0$, and $\eta > 0$, defined below:

– $\gamma \geq 0$ is the parameter, which positions the curve according to its origin. We call this the localization parameter and it is homogenous in time.

– $\beta > 0$ is a positive adimensional form, as it transmits form according to normal, exponential, or log-normal distribution. For example, for $\beta = 3.7$, the Weibull distribution approaches Gaussian distribution.

– $\eta > 0$ is a positive parameter of scale. It expresses lifespan.

Mean: $\mu = \gamma + \eta \times \Gamma\left(1 + \frac{1}{\beta}\right) = \Gamma\left(\frac{\gamma + 1}{\gamma}\right)$

Coefficient of variation, CV, (Weibull):

$$CV = \sqrt{\Gamma\left\{\left(\frac{\gamma+2}{\gamma}\right) \Big/ \Gamma\left(\frac{\gamma+1}{\gamma}\right)^2\right\} - 1}$$

Variance: $V = \sigma^2 = \eta^2 \times \left\{\Gamma\left(1 + \frac{2}{\beta}\right) - \Gamma^2\left(1 + \frac{1}{\beta}\right)\right\}$

Standard deviation: $\sigma = \eta \times \sqrt{\Gamma\left(\frac{\gamma+2}{\gamma}\right) - \Gamma^2\left(\frac{\gamma+2}{\gamma}\right)}$

Third-order moment of the random variable *X*:

$$S_x^3 = \eta^3 \times \Gamma\left(1 + \frac{3}{\beta}\right) + 3\gamma \cdot \eta^2 \times \Gamma\left(1 + \frac{2}{\beta}\right) + 3\gamma \cdot \eta^2 \times \Gamma\left(1 + \frac{1}{\beta}\right) + \gamma^3 \qquad [1.49]$$

Law **Γ(x)** is a Eulerian function of the second order (type). It is tabulated and the table of these values *a* and *b* according to β is presented in the Appendix of Volumes 2 and 3 of this series:

$$\left\{ a = \Gamma\left(1 + \frac{1}{\beta}\right) \text{ and } b = \sqrt{\Gamma\left(1 + \frac{2}{\beta}\right) - \Gamma^2\left(1 + \frac{1}{\beta}\right)} \right\} \qquad [1.50]$$

As this Eulerian function depends on β:

$$\mu = \gamma + a \times \eta \text{ and } \sigma = b \times \eta \quad [1.51]$$

Probability density:

$$f(\tau) = \frac{\beta}{\eta} \times \left(\frac{\tau - \gamma}{\eta}\right)^{\beta - 1} E\,\mathrm{xp}\left[-\left(\frac{\tau - \gamma}{\eta}\right)^{\beta}\right] \text{ for } \tau \geq \gamma;\ f(\tau) = 0 \text{ with } \tau \prec \gamma \quad [1.52]$$

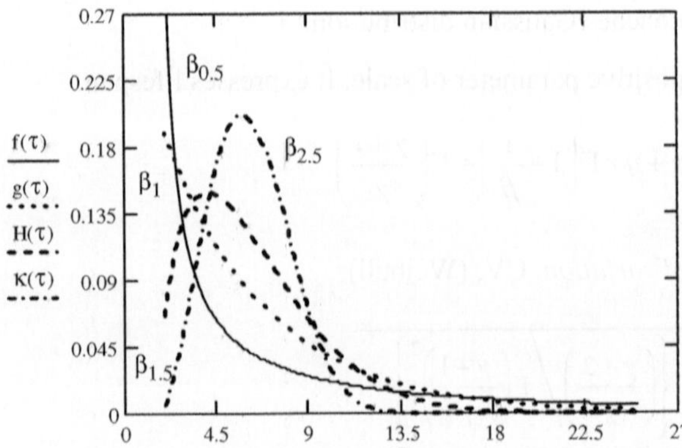

Figure 1.19. *Probability density of the Weibull distribution for three parameters*

CDF F(τ):

$$F(\tau) = 1 - E\,\mathrm{xp}\left[-\left(\frac{\tau - \gamma}{\eta}\right)^{\beta}\right] \text{ for } \tau \geq \gamma \text{ and } \mathrm{F}(\tau) = 0 \text{ with } \tau \prec \gamma \quad [1.53]$$

Failure rate, Z(τ):

$$Z(\tau) = \left(\frac{\beta}{\eta}\right) \times \left(\frac{\tau - \gamma}{\eta}\right)^{\beta - 1} \text{ for } \tau \geq \gamma \text{ and } \mathrm{Z}(\tau) = 0 \text{ with } \tau \prec \gamma \quad [1.54]$$

Reliability, R(τ):

$$R(\tau) = E\,\mathrm{xp}\left[-\left(\frac{\tau - \gamma}{\eta}\right)^{\beta}\right] \text{ for } \tau \geq \gamma \text{ and } \mathrm{R}(\tau) = 1 \text{ with } \tau \prec \gamma \quad [1.55]$$

We notice here that the Weibull distribution is delicate, especially when manipulating it with three parameters. It is often used in maintenance, but we propose using the Allan Plait graph presented in the Appendices, Volume 2 and 3 (Figure A.2).

1.12. The Weibull distribution (with two parameters)

In the case of γ =0, or when we change a variable, we obtain the Weibull distribution with two parameters (β and η). The mean and the variance (or the standard deviations) are calculated using interactions such as the Eulerian function $\Gamma(x)$, (see Appendices, Volumes 2 and 3, Table A.1).

Probability density, f(τ):

$$f(\tau)=\frac{\beta}{\eta}\times\left(\frac{\tau}{\eta}\right)^{\beta-1} E\text{xp}\left[-\left(\frac{\tau}{\eta}\right)^{\beta}\right] \textit{for } \eta\succ 0 \text{ and } \beta\succ 0 \qquad [1.56]$$

For β_1 = ½; β_2 = 1.0; β_3 = 1.5; β_4 = 2.5; η = 5, and τ = [0 to 25]

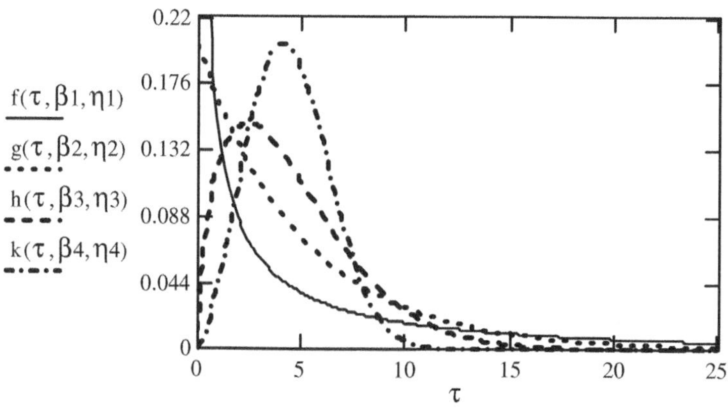

Figure 1.20. *Probability density of the Weibull distribution with two parameters*

Failure rate:

$$Z(\tau)=\lambda(\tau)=\left(\frac{\beta}{\eta}\right)\times\left(\frac{\tau}{\eta}\right)^{\beta-1} \quad \textit{for } \eta\succ 0 \textit{ and } \beta\succ 0 \qquad [1.57]$$

CDF (or cumulative probability) and inverse cumulative distribution (or inverse cumulative probability):

$$F(\tau) = 1 - E\,\mathrm{xp}\left[-\left(\frac{\tau}{\eta}\right)^{\beta}\right] \quad \textit{for } \eta \succ 0 \text{ and } \beta \succ 0 \qquad [1.58]$$

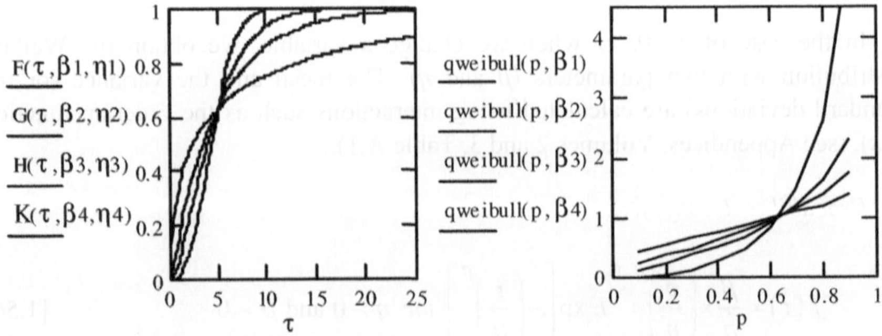

Figure 1.21. *Cumulative distribution function of the Weibull distribution with two parameters*

Reliability:

$$R(\tau) = 1 - F(\tau) = E\,\mathrm{xp}\left[-\left(\frac{\tau}{\eta}\right)^{\beta}\right] \quad \textit{for } \eta \succ 0 \text{ and } \beta \succ 0 \qquad [1.59]$$

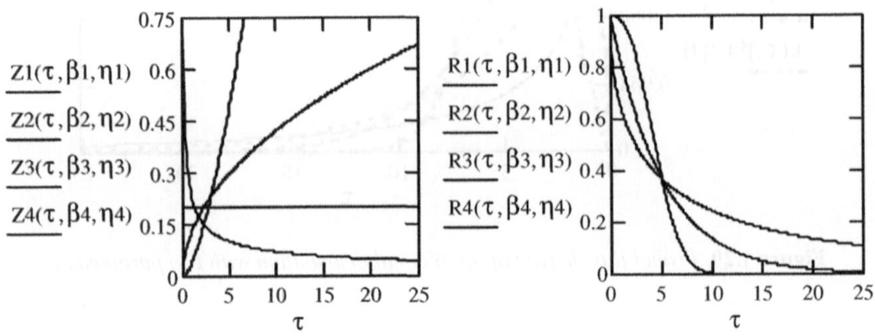

Figure 1.22. *Failure rate Z(τ) and reliability R(τ) of the Weibull distribution (β, η)*

From the example in the Appendices, we can see the graph on a functional scale (Allan Plait graph Figure A.1). Refer to Chapter 3 for parameter estimators (β and η) for the Weibull distribution with two parameters. The random variables *rWeibull* (m, β) give the Weibull distribution trace with two parameters:

$$rWeibull(m,\beta_1) = \{\ 1.078;\ 0.721;\ 49.587;\ 0.066;\ 5.808;\ 0.144;\ 3.498\ \}$$

$$rWeibull(m,\beta_2) = \{\ 3.985;\ 0.434;\ 0.090;\ 0.840;\ 1.416;\ 0.425;\ 3.779\ \}$$

$$rWeibull(m,\beta_3) = \{\ 0.709;\ 0.839;\ 1.130;\ 1.047;\ 0.628;\ 0.063;\ 1.529\ \}$$

$$rWeibull(m,\beta_4) = \{\ 0.578;\ 0.743;\ 0.675;\ 0.797;\ 0.816;\ 0.144;\ 0.871\ \}$$

The analyses of industrial incidents that have occasionally occurred allow us to trace distributions of material reliability. In such laboratories, not all trials carried out on weak structures (particularly aero-spatial structures) are conducted by taking every material involved to its breaking point. This is also true in certain areas of medicine (biomechanical medicine) where materials that do not fail are taken away at unexpected moments during their service life.

As rates of failure are often attributed to time, we use the Weibull distribution with two or three parameters. Reliability treatments are diverse and in training colleges, these techniques are developed using graphical adjustment resulting in parameter estimators. It should be noted that punctual parameter estimation remains insufficient if we wish to describe the law of survival correctly, as the weaker the population, the more uncertainty there is. For this reason, we have taken parameter confidence intervals from survival law into consideration. Unfortunately, in the case of the Weibull distribution, the calculation of confidence intervals is sometimes delicate, even difficult to lead due to the specificity of this law.

1.12.1. *Description and common formulae for the Weibull distribution and its derivatives*

The distribution of extreme values (Gumbel or VE) usually refers to the distribution of the minimum of a large number of unlimited random observations. We have already mentioned EV distributions when describing the uses of the Weibull distribution. From these extreme distributions of value come the distributions of limitation for the minimum or the maximum of a large collection of data from random observations of the same arbitrary distribution. Gumbel [GUM 54] showed that no matter what sort of initial, simple distribution (i.e. $F(x)$ which is continuous and derivable), only a few models are necessary, according to the interest held on the maximum of the minimum and whether the observations are limited.

The distributions of extreme values are common to the minimum reliability. For example, if a system consists of identical components *n* in series and the system is failing when the first of these components is faulty, the system's time-to-fail is the minimum of the time-to-fail of the random components *n*. Extreme value theory says that, independently of the choice of the model of the component, the model of the system will approach the Weibull distribution when *n* becomes considerably high. The same reasoning can also be applied to the simple level of a component, if the damage of the component occurs when the first of many similar components interact (connect together) and if the process of failure reaches a critical level. A certain distribution, the study of which we always come back to, is a distribution called extreme value, which is, in fact, the distribution of a minimal limitation of a high number of unlimited random variables, identically distributed and represented by the probability density function and the CDF.

– *Probability density f(x)*:

$$f(x) = \frac{1}{\beta} \times E\,xp^{\left(\frac{x-\mu}{\beta}\right)} \times E\,xp^{-E\,xp\left(\frac{x-\mu}{\beta}\right)} \quad for -\infty \prec x \prec +\infty \text{ and } \beta \succ 0 \qquad [1.60]$$

– *CDF F(x)*:

$$F(x) = 1 - E\,xp^{-E\,xp\left(\frac{x-\mu}{\beta}\right)} \quad for -\infty \prec x \prec +\infty \text{ and } \beta \succ 0 \qquad [1.61]$$

For *f*(x), $\beta = 2$; for g(x) $\beta_1 = 1$; and for *h*(x), $\beta_2 = 2$.

h(x) =

	0
0	$4.521 \cdot 10^{-6}$
1	$5.522 \cdot 10^{-6}$
2	$6.744 \cdot 10^{-6}$
3	$8.237 \cdot 10^{-6}$
4	$1.006 \cdot 10^{-5}$
5	$1.229 \cdot 10^{-5}$
6	$1.501 \cdot 10^{-5}$
7	$1.833 \cdot 10^{-5}$
8	$2.239 \cdot 10^{-5}$
9	$2.735 \cdot 10^{-5}$
10	...

$$h(x) = \left(\frac{1}{\beta 2}\right) \cdot e^{\left(\frac{x-\mu}{\beta 2}\right)} \cdot e^{-e^{\left(\frac{x-\mu}{\beta 2}\right)}}$$

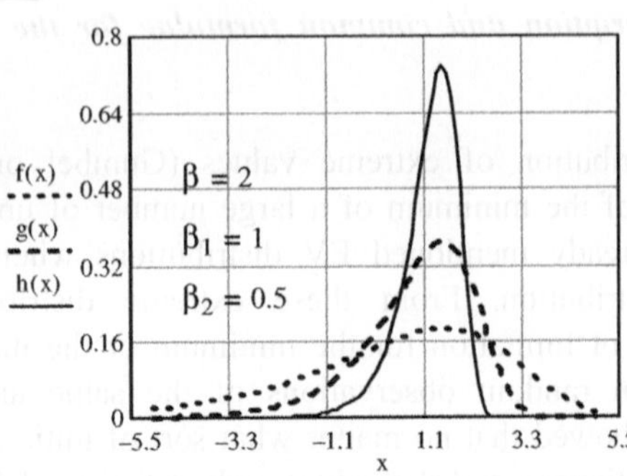

Figure 1.23. *Function of density of probability of the distribution of the extreme value (feature of the Weibull distribution with two and three parameters)*

For F(x) β = 2 ; For G(x) β_1 = 1 and for H(x) β_2 = 0.5

F(x) =

	0
0	0.038
1	0.04
2	0.042
3	0.044
4	0.046
5	0.049
6	0.051
7	0.054
8	0.056
9	...

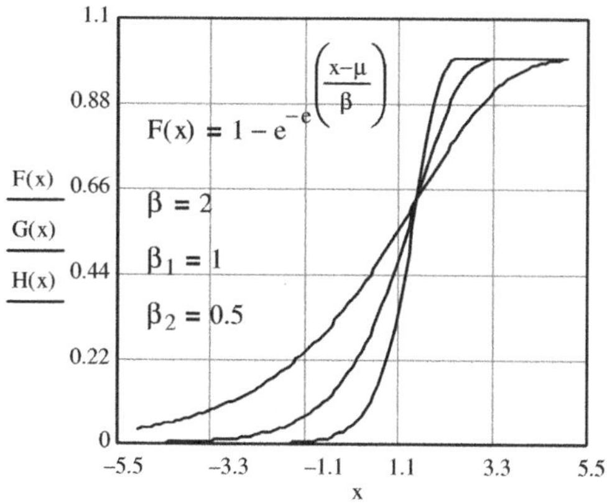

Figure 1.24. *Cumulative distribution functions of probability of the distribution of the extreme value (feature of the Weibull distribution with two and three parameters)*

For F(x) $\beta = 2$; for G(x) $\beta_1 = 1$; and for H(x) $\beta_2 = 0.5$.

The following are the uses of the extreme value distribution model (Gumbel as a development of the Weibull distribution):

1) Due to the flexible form and capacity of the model and a wide range of rates of failure, the Weibull distribution has been successfully used for many different applications as a purely empirical model.

2) Weibull's model can theoretically be taken as a distribution function of extreme value, held on the occurrence of "the weakest linking of the components of a given structure". This may explain the enthusiasm among scientists (not necessarily among mathematicians) for using it for many different purposes, from lifespan calculations to calculating the rolling of marbles, to calculations involving materials and structures.

3) The special case of the Weibull distribution takes (β) as its form parameter, equal to two. This distribution is known as the Rayleigh distribution. This final theoretical probability model is used to achieve a magnitude of radial forces.

1.12.2. *Areas where the extreme value distribution model can be used*

The EV distribution model (Gumbel distribution) has many industrial and technical applications. It is used in the following:

– Any modeling application where the variable is the minimum of many random factors (positive or negative values). For the modeling of functional lifespan since the time-to-fail is limited by zero. All of this indicates that Weibull's distribution proves to be the best choice.

– Weibull's distribution and extreme value distribution, which have a useful mathematical relationship. If τ_1, τ_2,..., τ_n is a scale of random times-to-fail of a Weibull-like distribution, ln τ_1, ln τ_2, ..., ln τ_n are random observations of extreme value distribution. To put it another way, the natural logarithm of the random time of Weibull is an extreme value of random observation.

If the Weibull distribution has a given parameter of form γ and α is a parameter that characterizes lifespan, the EV distribution at the scale of the natural logarithm has a mean (μ) of: $\mu = Log(\alpha)$ and $\beta = (1/\gamma)$.

1.12.2.1. *Extreme value pitch*

If the component or the system happens to fail when the first of numerous failure processes reaches a critical point, the extreme value theory suggests that the Weibull distribution is a good model. The central limit theorem advocates Gaussian distributions in the engineering field when the observed measures result in the accumulation of many random sources (such as measurement errors). Practical experience supports this theory. It is less well known that the extreme value theory suggests the possibility of Weibull's theory modeling failure time.

1.12.3. *Risk model*

We will now further develop this theme while looking at specific case studies (see Chapters 1 and 2 of Volume 2). The risk model intervenes when failure mechanisms are independent and the first failure of the component leads to the failure of the mechanism. The risk model is used to evaluate the reliability of the structure by the "set-up and layout" of the reliability models for each failure mode. The following conditions are necessary so that:

1) each failure mechanism leads to a particular type of failure, independent of the preceding one, until the appearance of another true fault occurs;

2) the component is said to have failed when the mechanisms of failure have failed;

3) each of the failure modes (κ) have a model of known life distribution $F_i(\tau)$.

The model of failure can be used when the three conditions are reunited. If $R_c(\tau)$, $F_c(\tau)$, and $h_c(\tau)$ denote, respectively, reliability, the CDF, and the failure rate of the component, and $R_i(\tau)$, $F_i(\tau)$, and $h_i(\tau)$ represent reliability, the CDF, and the failure rate of the *i*th failure mode, respectively, then the risk model formulae are:

– *Product of reliability*:

$$R_c(\tau) = \prod_{i=1}^{k} R_i(\tau) \quad and \quad R_c(\tau) = 1 - \prod_{i=1}^{k}\left[1 - F_i(\tau)\right] \qquad [1.62]$$

– *Addition of the failure rate*:

$$h_c(\tau) = \sum_{i=1}^{k} h_i(\tau) \qquad [1.63]$$

All of the failure mechanisms have the possibility of failing.

1.12.4. *Products of damage*

The log-normal model, for example, can be applied when the degradation (damage) is caused by *random shocks* which increase degradation to a rate proportional to the total rate already present. That is, $y_1, y_2, \dots y_n$, which come from measurements taken to show the *extent of damage* of a particular process of successive failure. These measurements were taken at discrete time windows during which the process progressively reaches failure stage.

$$y_i = (1 + \varepsilon_i) \times y_{i-1} \qquad [1.64]$$

where the ε_i are, there are small cases of damage, independent of random disturbances or system shocks, which lead to failure. In other words, there is a rise in the total extent of degradation that is already present. This is what we call *multiplying damage (or the product of damage* due to the effect of consecutive training). We can express the total quantity of damage from the *n*th to the momentary as follows:

$$x_n = \left(\prod_{i=1}^{n}(1 + \varepsilon_i)\right) \times x_0 \qquad [1.65]$$

where x_0 is a constant and ε_i are little random shocks. Next, we take the natural logarithms of the two terms to obtain the following:

$$\ln x_n = \sum_{i=1}^{n} \ln(1 + \varepsilon_i) + \ln x_0 \approx \sum_{i=1}^{n} \varepsilon_i + \ln x_0 \qquad [1.66]$$

When using the central limit theorem argument, we can conclude that ln x_n is approximately a Gaussian distribution. But in log-normal distribution, that implies that x_n (or the rate of damage) will allow everything (at any moment τ) to follow a log-normal model. The failure mechanisms, which would follow a model of multiplying damage, are:

1) the growing or propagation of fissures;

2) chemical reactions leading to the formation of new composites;

3) the diffusion or migration of ions (atoms in physical metallurgy).

Numerous modes of failure, or modes used by semiconductors, are caused by one of three processes of degradation (damage) according to the wear and tear of the failure mechanisms due to the *growing of fissures*, to *corrosion*, *etc.*

1.13. The Birnbaum–Saunders distribution

The Birnbaum–Saunders distribution [BIR 69] was introduced to characterize the statistical distribution of time-to-fail in material fatigue. It is applicable in the case of damage caused by cyclical load. The probabilistic model derives from the random propagation of fissures arising over the course of numerous independent, repeated cycles of the *degradation* of the constraints leading to destruction. A key hypothesis is that the failure rate throughout a cycle is independent of failure within other cycles, with the same random distribution.

When this hypothesis corresponds well to a hypothetical physical model describing the process of this damage, we start using the Birnbaum–Saunders model (difference between the derivation of the log-normal model and the hypothesis of the lifespan in fatigue as well as Miner's rule). For the random variable τ that often represents the length of the fissure; η a scale parameter; γ a form parameter; and $\Phi(.)$ a standard tabulated function of the CDF $F(\tau)$, we propose the main characteristics of probabilistic modeling:

The *mean* μ presents itself as:

$$\mu = \left\{ \gamma \left(1 + \frac{\gamma^2}{2} \right) \right\}$$

Variance:

$$\sigma^2 = \eta^2 \cdot \gamma^2 \left(1 + \frac{5\gamma^2}{4} \right)$$

Standard deviation:

$$\sigma = \eta \cdot \gamma \sqrt{\left(1 + \frac{5\gamma^2}{4}\right)}$$

Probability density function:

$$f(\tau) = \left(\frac{1}{2\sqrt{\pi} \times \gamma^2 \times \eta \cdot \tau^2}\right) \times \frac{\tau^2 - \eta^2}{\sqrt{\frac{\tau}{\eta}} - \sqrt{\frac{\eta}{\tau}}} \times Exp\left\{-\frac{1}{\gamma^2} \times \left(\frac{\tau}{\eta} + \frac{\eta}{\tau} - 2\right)\right\} \quad [1.67]$$

CDF of probability:

$$F(\tau) = \Phi\left\{\left(\frac{1}{\gamma}\right) \cdot \left(\frac{\tau}{\eta} + \frac{\eta}{\tau}\right)\right\} \quad [1.68]$$

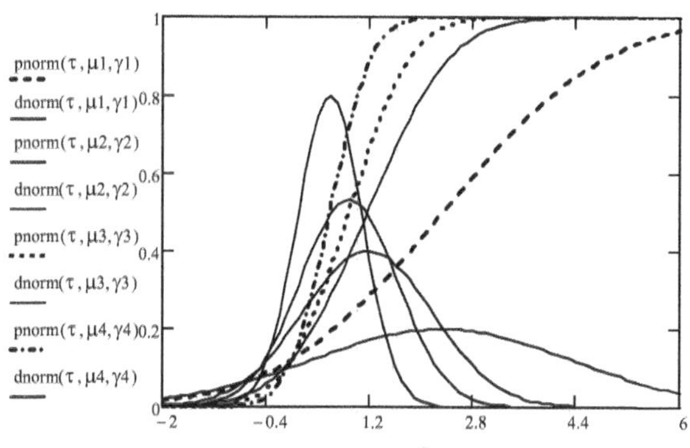

Figure 1.25. *Functions of lifespan probability distribution of a structure assembled by welding an interval of time τ = [1 to 30000] hours for the parameters of scale η = 1000 and shape parameters: γ_1=2; γ_2=1; γ_3= 0.75; and γ_3=0.5*

The probability distribution function is sometimes "extremely" biased (lengthened bias) for small gamma values (γ) as the above illustrates. If the fissure of each cycle of constraints is a random independent quantity of anterior cycles of increase, the model of the life mode during fatigue can be applied.

1.13.1. *Derivation and use of the Birnbaum–Saunders model*

It is a model that undergoes load cycles under permanent constraints.

Each cycle involves a dominant fissure propagating itself toward a critical length until it ends up causing failure. Under the application of repeated n cycles of load, the total extension of the dominant fissure can be written as:

$$g_n(a) = \sum_{j=1}^{n} \Psi_j \qquad [1.69]$$

We postulate that Ψ are independent and identically distributed non-negatively of the random variable μ average and the variance σ^2. We postulate that failure occurs at the N cycle when $g_n(a)$ passes a critical value, $g(a)$. If n is high, we can use the *central limit theorem* to deduce the following:

$$P_r\{N \leq n\} = 1 - P_r\left\{\sum_{j=1}^{n} \Psi_j \leq g_n(a)\right\} = \Phi\left\{\frac{\mu \cdot \sqrt{n}}{\sigma} - \frac{g_n(a)}{\sigma \cdot \sqrt{n}}\right\} \qquad [1.70]$$

As there are many cycles, each one takes place during a very short time. We can replace the number of discrete N cycles necessary to reach failure during the continuous time τ_f necessary to reach failure. The cumulative function of probability $F(t)$ of τ_f is given by the following expression:

$$F(\tau) = \Phi\left\{\left(\frac{1}{\alpha}\right) \times \left(\frac{\tau}{\beta} - \sqrt{\frac{\beta}{\tau}}\right)\right\} \text{ with } \beta = \left(\frac{g(a)}{\mu}\right) \text{ and } \alpha = \left(\frac{\sigma}{\sqrt{\mu \cdot g(a)}}\right) \qquad [1.71]$$

$\Phi(\,)$ is the value tabulated by log-normal distribution. The model written with parameters α and β is another way of writing the Birnbaum–Saunders distribution, often used as $\alpha = \gamma$ and $\beta = \mu$. Mainly, the hypothesis in derivation, from a physical point of view, is the increase in fissures throughout the course of a cycle. Moreover, the increase has approximately the same random distribution from cycle-to-cycle. It is a different situation from the argument of proportional damage used to derive a model of log-normal distribution, with failure rate at whatever point according to total rate of damage, which occurs at that particular instant.

This type of physical damage is in accordance with Miner distribution (see Chapter 3). Birnbaum's hypothesis postulates that the system is physically restrictive. It conforms more to the deterministic model of material physics known as Miner's rule (rule implying damages which arise after n cycles, at a level of constraints which produces a lifespan in fatigue of N cycles and proportional to n/N). Therefore, when the physics of destruction suggests the application of Miner distribution, the Birnbaum–Saunders model proves to be more appropriate to a model of distribution of lifespan structure.

1.14. The Cauchy distribution

Cauchy distribution does not have a mean or a standard deviation. Even without skewness and kurtosis, this is a Cauchy distribution (sometimes called Lorentz's law in physics) and it somewhat resembles normal distribution. However, the curves have lines that are relatively longer than Gaussian "bells". In mechanical reliability, the tests carried out on the data distributed according to the Cauchy distribution constitute an excellent indicator of sensitivity.

Furthermore, Cauchy distribution allows the integration of a considerable variety of distribution hypotheses. The mean and standard deviation of the Cauchy distribution are undefined. The practical meaning can be explained by the fact that the collection itself, made up of 77,000 points of data, does not allow a more precise estimation of the mean and difference that will make seven points. Here are the statistical characteristics of this law:

– Mean:	The mean is undefined
– Median:	Localization τ
– Mode:	The localization parameter τ
– Range:	$\{-\infty$ to $+\infty\}$
– Standard deviation:	The standard deviation is undefined
– Coefficient of variation (CV):	The coefficient of variation is undefined
– Coefficient of skewness:	*Skewness is undefined*
– Coefficient of kurtosis:	*Kurtosis is undefined*

1.14.1. *Probability density function*

Cauchy distribution is expressed in the following way by probability density:

$$f(\tau,\eta,\gamma)=\left(\frac{1}{\pi\gamma}\right)\times\left(1+\left(\frac{\tau-\eta}{\gamma}\right)\right)^{-1} \quad [1.72]$$

where η is the localization parameter and γ the scale parameter.

We have often turned to scale parameters and localization parameters, even shape parameters (for the Weibull distribution to three parameters) in different modeling applications. For example, a null localization parameter ($\gamma = 0$) with a unitary scale parameter ($\eta = 1$) characterizes a normal standard distribution N [0.1].

With N [0.1], i.e. ($\gamma = 0$) and ($\eta = 1$) this is called Cauchy *standard distribution*, hence the mathematical expression:

$$f(\tau) = 1 \Big/ \left[\pi\left(1+\tau^2\right)\right] \qquad [1.73]$$

The localization parameter ($\gamma = 1.5$) has a visual effect, allowing easier reading of the normal-distribution graph on τ [–6, +6]. A localization parameter –6 would have exceeded six units to the right on the τ axis, etc.

Probability distribution function of cumulative probability:

$$F(\tau,\eta,\gamma) = \left(\frac{1}{\pi}\right) \times \arctan\left(1 + \left(\frac{\tau-\eta}{\gamma}\right)\right) + \frac{1}{2} \qquad [1.74]$$

The formula for Cauchy cumulative probability distribution is:

$$F(\tau) = \left(\frac{1}{2}\right) + \left(\frac{\arctan(\tau)}{\pi}\right) \qquad [1.75]$$

The Cauchy *probability density function* is plotted like this:

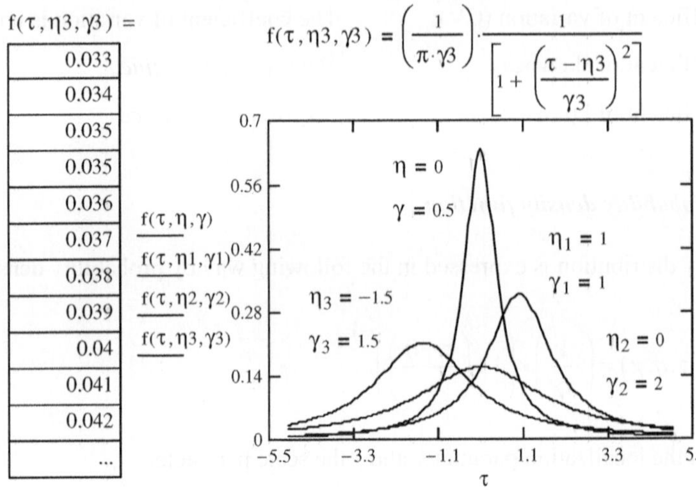

Figure 1.26. *Graph showing Cauchy probability density*

The Cauchy CDF is graphically plotted as in Figure 1.27.

$F(\tau,\eta,\gamma) =$
-0.309
-0.309
-0.308
-0.308
-0.308
-0.307
-0.307
-0.307
-0.306
-0.306
...

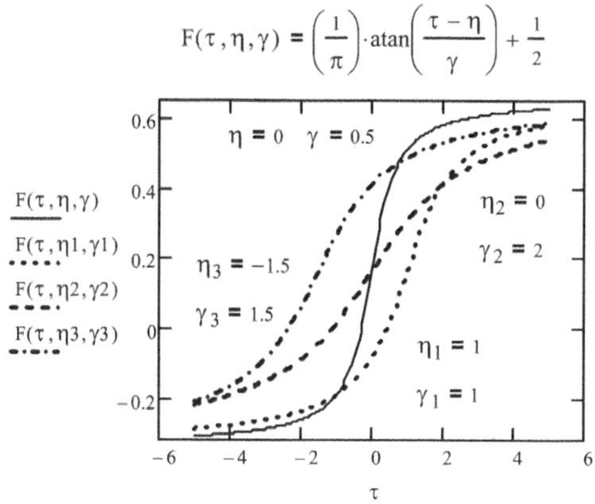

Figure 1.27. *Graphs showing the Cauchy cumulative distribution function*

The Cauchy distribution function (*percentile*) is given as:

$$G(\tau) = -\left(\frac{1}{F(\tau,\eta,\gamma)}\right) \tag{1.76}$$

The *standard function* is written as:

$$G(\tau) = -\cot(\pi\tau) = -\left(\frac{1}{\tan(\pi\tau)}\right) \tag{1.77}$$

The Cauchy *inverse cumulative distribution* is graphically plotted below:

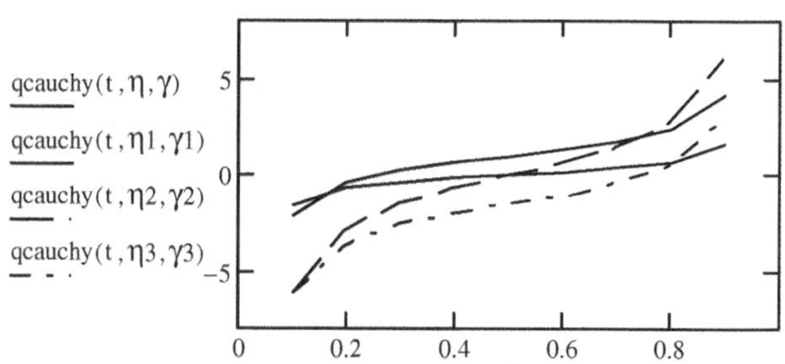

Figure 1.28. *Graphs showing Cauchy percentile function*

1.14.2. *Risk function*

The Cauchy risk function is calculated from Cauchy probability density and the distribution function of cumulative probability. The function we call **risk** is, in fact, the relationship between function and distribution and cumulative distribution. It is presented as follows:

$$r(\tau,\eta,\gamma)=\frac{f(\tau,\eta,\gamma)}{S(\tau,\eta,\gamma)}=\frac{f(\tau,\eta,\gamma)}{1-F(\tau,\eta,\gamma)}=\frac{\left(\dfrac{1}{\pi\gamma}\right)\times\dfrac{1}{1+\left(\dfrac{\tau-\eta}{\gamma}\right)}}{1-\left(\dfrac{1}{\pi}\times\arctan\left(1+\left(\dfrac{\tau-\eta}{\gamma}\right)\right)+\dfrac{1}{2}\right)} \quad [1.78]$$

The *risk* function and the function of the Cauchy cumulative probability (chance) are graphically plotted as in Figure 1.29.

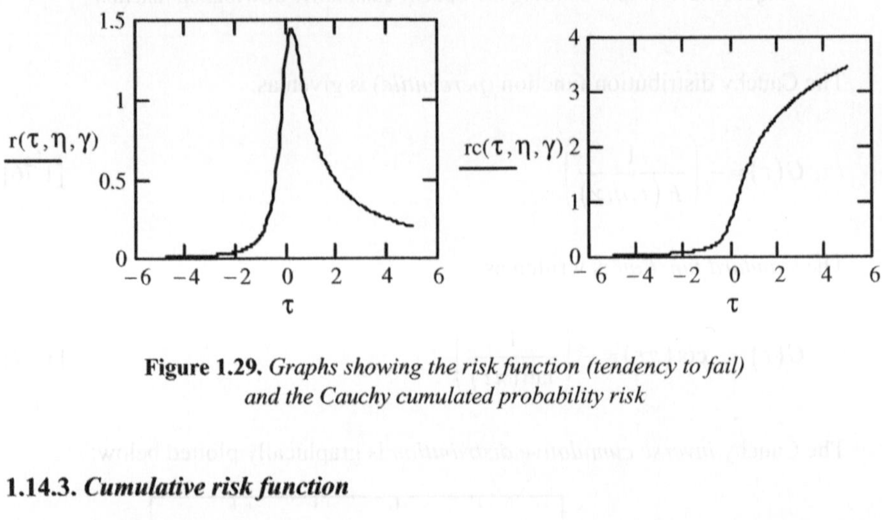

Figure 1.29. *Graphs showing the risk function (tendency to fail) and the Cauchy cumulated probability risk*

1.14.3. *Cumulative risk function*

The function said to be by chance (Cauchy) is calculated by the integral of the risk function being able to thus be interpreted as momentary failure probability (T) of survival until moment (τ):

$$r_c(\tau)=\int_{-\infty}^{\tau} r(\tau)d(t) = -\ln\{1-F(\tau,\eta,\gamma)\} \quad [1.79]$$

1.14.4. *Survival function (reliability)*

In mechanical reliability, this function is widely used. It is in fact probability which takes the random variable (T) and which is superior to (τ). The Cauchy survival function (or its very reliability) is calculated as follows:

$$R(\tau) = P_r\{T \succ \tau\} = \{1 - F(\tau,\eta,\gamma)\} \qquad [1.80]$$

1.14.5. *Inverse survival function*

Following the example of Cauchy percentile function, the Cauchy inverse survival function is calculated by the inversion of the CDF. The graph and mathematical expression of this is shown as:

$$R_{inv}(\tau,\eta,\gamma) = \frac{1}{F(\tau,\eta,\gamma)} \qquad [1.81]$$

Vector of ***m*** random numbers of a Cauchy distribution. For $m = 7$, we will have the following when t varies from (0 to 1]:

$$rCauchy(m,\eta,\gamma) = \{\ 1.619;\ \ -0.072;\ \ 1.150;\ \ -4.046;\ \ 2.457;\ \ -0.877;\ \ -1.299\ \}$$

$$rCauchy(m,\eta_1,\gamma_1) = \{\ -1.564;\ \ 1.257;\ \ 1.890;\ \ 1.590;\ \ 2.460;\ \ 127.847;\ \ -0.337\ \},\ etc.$$

The *survival* functions of Cauchy and Cauchy *inverse survival* are graphically plotted below:

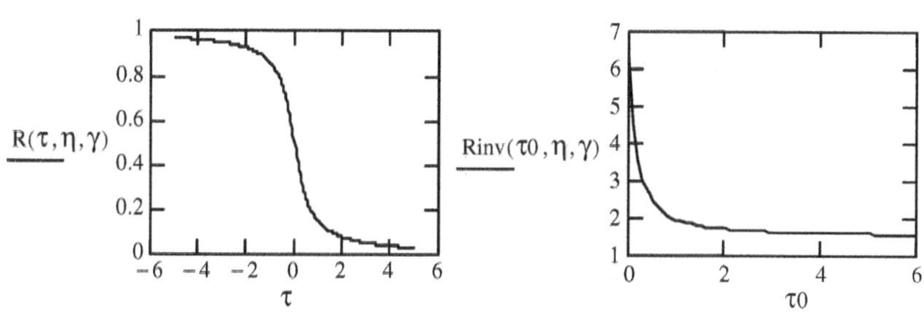

Figure 1.30. *Graphs showing Cauchy survival functions and Cauchy inverse survival functions*

1.15. Rayleigh distribution

We owe this distribution to Lord Rayleigh. It is used in the analysis of spectral densities and is perfectly adapted to the reliable study of materials and structures when the approach of the behavior of structures is analyzed under the angle of undulations (wind, sea). Originally, it has always been used mostly for thermodynamics in the modeling of particles. It is essentially characterized by its standard deviation, a real positive $\sigma > 0$ for a variable $\tau \in [0, \infty)$. When the coordinates are independent, centered, and admit the same variance in a bi-dimensional Gaussian distribution, the Rayleigh distribution intervenes in the description of the envelope of a narrowband Gaussian process (noise when leaving receptors).

The *mean* is written as $\sigma \times \sqrt{\pi/2}$ and the *mode* as σ.

The *variance* takes the form:

$$\left(4 - \frac{\pi}{2}\right) \times \sigma^2$$

and the *median*: $\sqrt{Ln(4)} \times \sigma$.

Skewness: $\dfrac{2\sqrt{\pi}\,(\pi - 3)}{\sqrt[3]{(4-\pi)}}$;

Kurtosis: $-2\left(\dfrac{3\pi^2 - 12\pi + 8}{(4-\pi)^2}\right)$

Density (function) of the Rayleigh distribution:

$$f(\tau) = \left(\frac{\tau}{\sigma^2}\right) \times E\,\mathrm{xp}\left(-\frac{\tau^2}{2\sigma^2}\right) \qquad [1.82]$$

This is how this function appears *for: $\sigma_1 = 1$; $\sigma_2 = 1.5$; $\sigma_3 = 2$; $\sigma_4 = 3$.*

To estimate this function, we introduce a number (κ), which is an independent Rayleigh variable but with the same parameter law (σ). The estimator said to be the most plausible of (σ) appears below:

$$\hat{\sigma} = \sqrt{\left(\frac{1}{2\kappa}\right) \times \sum_{i=1}^{\kappa} \tau_i^2} \qquad [1.83]$$

CDF:

$$F(\tau) = 1 - E\mathrm{xp}\left(-\frac{\tau^2}{2\sigma^2}\right) \qquad [1.84]$$

$f(\tau, \sigma 1) =$

0
0.1
0.196
0.287
0.369
0.441
0.501
...

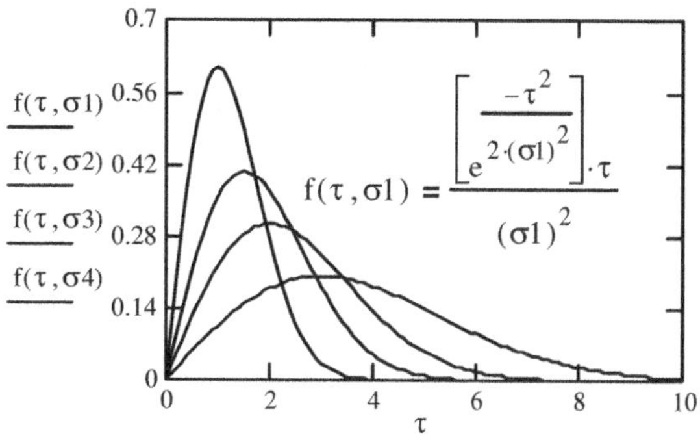

Figure 1.31. *Graph showing the probability density of Rayleigh distribution*

For $\sigma_1 = 1$; $\sigma_2 = 1.5$; $\sigma_3 = 2$; *and* $\sigma_4 = 3$, we will now trace the CDF as in Figure 1.32.

$F(\tau, \sigma 1) =$

0
$4.988 \cdot 10^{-3}$
0.02
0.044
0.077
0.118
0.165
...

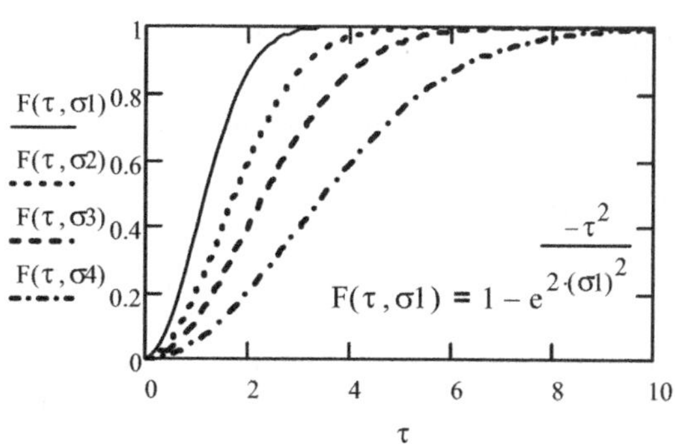

Figure 1.32. *Graph showing the Rayleigh cumulative distribution function*

1.16. The Rice distribution (from the Rayleigh distribution)

The Rice distribution is, in fact, a generalization of the Rayleigh continuous distribution. It is useful in the study of signal treatment. The parameters that characterize this function are usually (ν and σ) ≥ 0. The support of variable τ varies between [0 and $+\infty$). We have deliberately omitted the presentation of complications of skewness and the kurtosis of curves. Additionally, from the literature, we have two centered independent Gauss variables of equivalent variance (RV) = σ^2. If they represent two coordinates from a point on a map, the distance of this point from the origin allows us to write the Rayleigh distribution expression using the above formula.

Supposing that the distribution is centered on a point of coordinates [vcos (θ), vsin (θ)], in polar coordinates (v, θ), probability density takes the form:

$$f(\tau|\nu,\sigma)=\frac{1}{\sigma^2}\times \text{Exp}\left(\frac{-\left(\tau^2+\nu\right)^2}{2\sigma^2}\right)\times \text{I}_0\left(\frac{\tau\cdot\nu}{\sigma^2}\right) \qquad [1.85]$$

To solve this equation, we must first find the solution to the factor: $\text{I}_0\left((\tau\cdot\nu)/\sigma^2\right)$ of Bessel's function of the first order (modified hyperbolic function). Bessel's expression generally appears as:

$$\tau^2\left(\frac{\partial^2 y}{\partial\tau^2}\right)+\tau\left(\frac{\partial y}{\partial\tau}\right)+\left(\tau^2-\nu^2\right)y=0 \qquad [1.86]$$

where $\text{I}_0\left(\tau\cdot\nu/\sigma^2\right)$ is Bessel's function, modified by type 1 and by order 0, hence the resolution has been programmed using MathCAD $\text{I}_0(Z)$. We will present neither this function in detail, nor its links, but will only use the result of Bessel's factor I( ) as it comes into play in the Rice distribution function.

Probability density of *Rice* distribution:

For: $\sigma = 3/2$; $\nu_1 = 0.5$; $\nu_2 = 0.75$; $\nu_3 = 1.5$; $\nu_4 = 1.75$; *and* $\nu_4 = 2$.

The *CDF* for Rice distribution is graphically presented in a similar way to the Rayleigh distribution. It can sometimes take the modified form:

$$F(\tau|\nu,\sigma)=\left\{1-Q_1\left(\frac{\nu}{\sigma},\frac{\tau}{\sigma}\right)\right\}\ \textit{with}\ Q_1\left(\frac{\nu}{\sigma},\frac{\tau}{\sigma}\right)\ \textit{Marcum 's Q function} \qquad [1.87]$$

$I0\left(\frac{\tau \cdot v1}{\sigma^2}\right) =$	$f(\tau,\sigma,v1) =$
	0.000
1	0.043
1	0.086
1	0.129
1.001	0.170
1.002	0.211
1.003	0.250
1.004	0.288
...	...

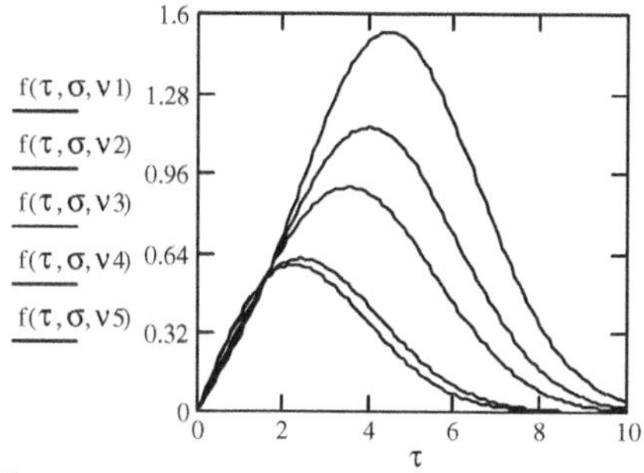

Etc. ...

Figure 1.33. *Graphs showing the Rice distribution function*

For: $\sigma_1 = 0.5$; $\sigma_2 = 0.75$; $\sigma_3 = 1.5$; $\sigma_4 = 1.75$; $\sigma_4 = 2$

$v_1 = 0.5$; $v_2 = 0.75$; $v_3 = 1.5$; $v_4 = 1.75$; $v_4 = 2$

$F(\tau,\sigma1,v1) =$
0
$8.849 \cdot 10^{-3}$
0.035
0.077
0.133
0.199
0.274
0.353
0.434
0.513
...

Etc. ...

$$F(\tau,\sigma1,v1) = 1 - e^{\frac{-\tau^2}{2\cdot(\sigma1)^2}}$$

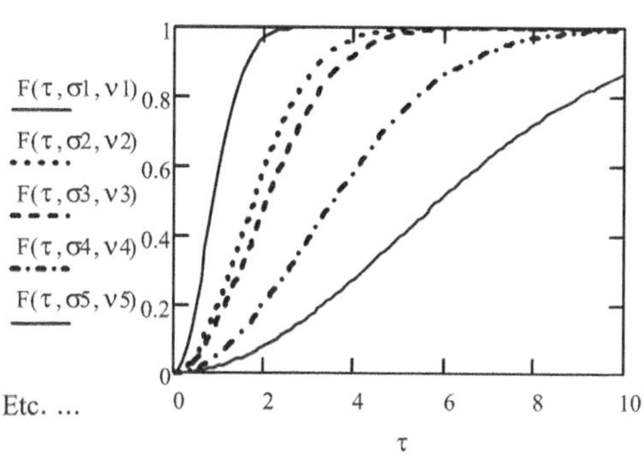

Figure 1.34. *Graph showing the Rice cumulative distribution function*

1.17. The Tukey-lambda distribution

The Tukey-lambda density function does not simply have a known form. It is for this reason, among others, that the use of this work essentially for reliability has never taken place. Tracing all of the curves for shape parameters (λ) would have the same effect as tracing five other curves mentioned below. The Tukey-lambda distribution law is known rather implicitly by the distribution of its quantiles. It is generally presented as:

$$T(\tau)=\frac{\tau^{\lambda}-(1-\tau)^{\lambda}}{\lambda} \qquad [1.88]$$

It is a law of distribution which, according to the values, would take the shape parameter (λ) in the function which would take the form:

It is similar to Cauchy's	*If* $\lambda=-1$
It is exactly a logistic law	*If* $\lambda=0$
It is approximately normal	*If* $\lambda=0.14$
It is similar to (concave) uniform law	*If* $\lambda=½$
It is normal distribution exactly	*If* λ *is situated between* (–1) and (+1)

Table 1.4. *Cauchy distribution in functions of* (λ)

The Tukey-lambda functions called Ramberg and Schmeiser are written as:

$$T(\tau)=\lambda_1+\left(\left(\tau^{\lambda_3}-(1-\tau)^{\lambda_4}\right)\Big/\lambda_2\right) \qquad [1.89]$$

The Tukey-lambda version named after Freimer, Mudholkar, Kollia, and Lin is written as:

$$T(\tau)=\lambda_1+\left(\left(\frac{\tau^{\lambda_3}}{\lambda_3}-\frac{(1-\tau)^{\lambda_4}}{\lambda_4}\right)\Big/\lambda_2\right) \qquad [1.90]$$

The Tukey-lambda distribution is also used because of the distribution of its quantiles. These are in fact the *essential* points taken at regular vertical intervals of a function of cumulative distribution of an RV. Certain quantiles have specialist names such as the following: 100 quantiles are called centiles or percentiles; 10 are deciles; five are quintiles; four are quartiles, and two is the median.

Quantiles are useful measures because they are less sensitive to elongated distributions or aberrant values. For example, let us take a random variable which follows a law of normal distribution. Any specific sample of this random variable will have approximately a 75% chance of being below the mean. This is due to the presence of a long exponential distribution line in positive values, which is absent in negative values. This is an illustrative example of the above. Using the MathCAD program, we have called the probability density function the *dnorm* of mean 2, i.e. *mean* = 2. Then, we have generated the cumulative distribution function *pnorm* for $y = 0$ to 1.

What can the size of (x) take before the bell curve reaches ½? As expected, $\mu = 2$ will be the mean.

– The *CDF* is written (in MathCAD)as:

$$qnorm\left(\frac{1}{2},2,2\right)=2$$

– *Distribution* function:

$$\int_{-25}^{25} xdnorm(x,2,1)\,dx = 2 \text{ and } pnorm(2,2,1)=\frac{1}{2}$$

The 75th, 90th, and 95th percentile of the normal distribution function of parameter N(2,1) are calculated below. The graphical interpretation of what precedes and what follows is presented below:

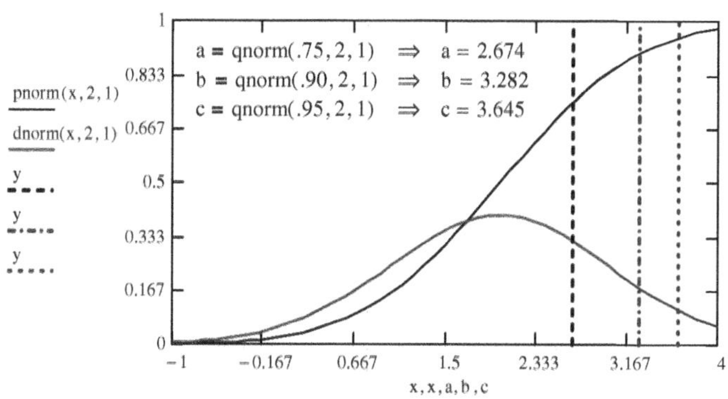

Figure 1.35. *Graph showing probability density and cumulative distributions of quantiles and of normal distribution*

1.18. Student's (t) distribution

Usually, Student's (t) distribution (see Appendices, Volumes 2 and 3, Table A.3) is used for the development of hypothesis tests and calculating confidence intervals. It almost never intervenes in applications of model establishment. (t) distribution is symmetrical in almost all cases. The statistical characteristics of this law are:

– Domain $\Re$;

– Mean (the mean is not defined for $\nu = 1$);

– Median = 0 and mode = 0.

Standard deviation: $\sigma = \sqrt{\dfrac{\nu}{\nu-2}}$. It is not defined for ν equal to 1 n_i 2;

Coefficient of variation: undefined; skewness = 0, is not defined for $\nu \leq 3$

Kurtosis: $K = \{3(\nu-2)/(\nu-4)\}$, not defined for $\nu \leq 4$.

Probability density function: We present probability density function as follows:

$$f(\tau,\nu) = \frac{\Gamma\left(\dfrac{\nu+1}{2}\right)}{\Gamma\left(\dfrac{\nu}{2}\right)\cdot\sqrt{\pi\nu}} \times \left(1+\frac{\tau^2}{\nu}\right)^{-\left(\frac{\nu+1}{2}\right)} = \frac{\left(1+\dfrac{\tau^2}{\nu}\right)^{-\left(\frac{\nu+1}{2}\right)}}{B\left(\dfrac{1}{2},\dfrac{\nu}{2}\right)\cdot\sqrt{\nu}} \qquad [1.91]$$

B( ) is the beta function (tabulated) and ν is a parameter of whole positive form. The formula for the beta function is presented as follows:

$$B(.) = beta(\alpha,\beta) = \int_0^0 \left(\tau^{\alpha-1} \times (1-\tau)^{\beta-1}\right) d\tau \qquad [1.92]$$

The (t) distribution is often considered to have neither scale parameters nor localization parameters. Usually, the mean and the standard deviation [μ, σ] are considered scale and localization parameters, respectively. We can graphically illustrate the probability distribution function as well as the *cumulative distribution function* from Student's (t) distribution.

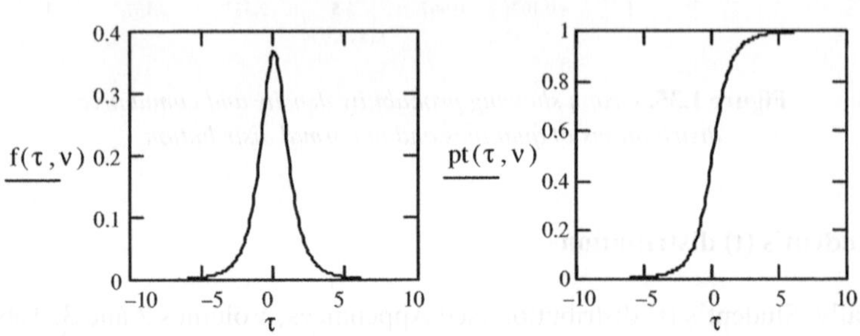

Figure 1.36. *Graphs showing the probability distribution of (T) and the cumulative distribution function of (t) distribution*

1.18.1. *t-Student's inverse cumulative function law (T)*

For m = 7

$$rt(m, v) = \begin{pmatrix} -0.71 \\ 0.903 \\ 0.372 \\ -1.194 \\ 1.124 \\ 0.696 \\ -0.63 \end{pmatrix}$$

etc. ...

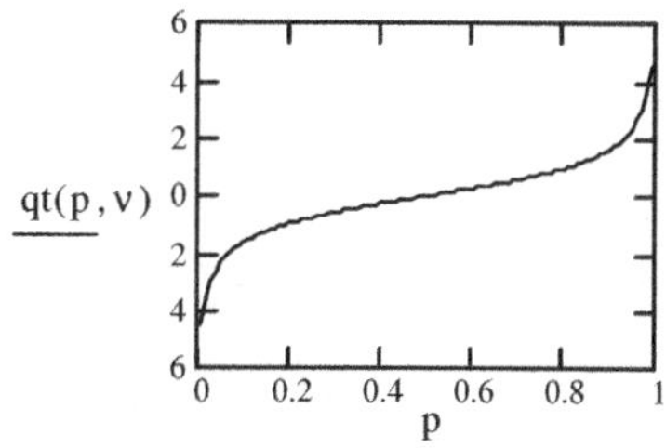

Figure 1.37. *Graph showing the inverse cumulative distribution function of (T) law. Vectors of random values following Student's distribution law (T)*

1.19. Chi-square distribution law (χ^2)

χ^2 distribution law (see Appendix, Table 7) is one of the most widespread in descriptive statistics, inherent to the development of hypothesis testing and confidence intervals. It is rarely used for applications of model establishment and so we will deliberately omit the estimation of parameters that characterize this law.

Domain: $\tau \in \{0\,;+\infty\}$

$$Mean = \mu\;; Median = \left\{\left(\frac{v-2}{3}\right) for\ the \uparrow numbers\right\}; \text{Mode} = (v-2)\, for\ v \succ 2$$

$$\text{standard deviation} = \sigma = \sqrt{2v}\;; kurtosis = \frac{\sqrt[3]{2}}{\sqrt{v}};\ skewness = \left(3+\frac{12}{v}\right)$$

1.19.1. *Probability distribution function of chi-square law (χ^2)*

$$\chi^2_{(\tau,v)} = \frac{Exp\left\{-\dfrac{\tau}{2}\right\}}{2\Gamma\left(\dfrac{v}{2}\right)} \times \left(\frac{\tau}{2}\right)^{\left(\frac{v}{2}-1\right)} \qquad [1.93]$$

where:

v is a positive integer (shape parameter) called the degrees of freedom of χ^2;

τ is a scalar or a vector (RV) of χ^2 distribution;

m is an integer > 0 of χ^2;

$\Gamma(.)$ is the gamma standard function (tabulated).

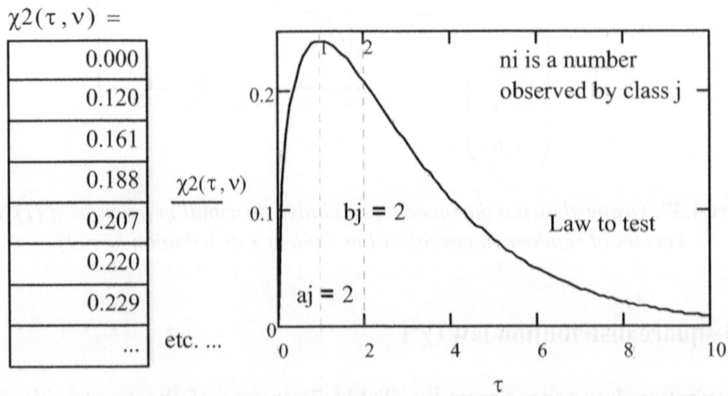

Figure 1.38. *Graph showing the probability distribution* ***dchisq(τ, ν)*** *of* χ^2 *law*

Because of these statistical tests, the distribution of χ^2 is considered "normalized distribution". It neither presents localization parameters nor scale parameters. Furthermore, in modeling, at any distribution, χ^2 law could take $N(\mu, \sigma)$ respective of localization parameter and scale.

1.19.2. *Probability distribution function of chi-square law (χ^2)*

Cumulative probability distribution function is written as:

$$F\left\{\chi^2(\tau,\nu)\right\} = \frac{\gamma\left(\dfrac{\nu}{2},\dfrac{\tau}{2}\right)}{\Gamma\left(\dfrac{\nu}{2}\right)} \quad for\ \tau \geq 0 \qquad [1.94]$$

where $\gamma(.)$ is the incomplete gamma function.

Inverse cumulative distribution *functions* (χ^2) of χ^2; dchisq (p, ν) distribution:

m, a whole positive > 0 is a vector of RV of χ^2; *rchisq* (m, ν) distribution

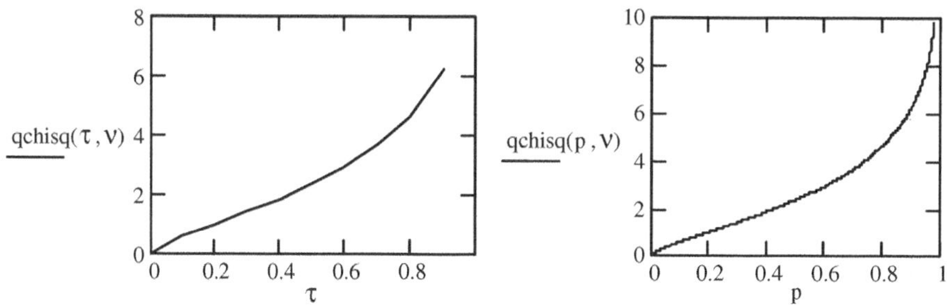

Figure 1.39. *Graphs showing the function of cumulative probability* ***qchisq(τ, ν)*** *of the law of inverse cumulative probability of χ^2 law*

For $m = 7$, we have $a = [7.381, 4.979, 0.610, 5.162, 2.611, 3.085, 7.739]$, etc.

The distribution of χ^2 law is very often used in cases where sections are critical for hypothesis testing and to also determine confidence intervals (see Appendices, Volumes 2 and 3, Tables A.7–A.9). In relation to this, we can give the example of the case inherent to χ^2 independence tests in a contingence table. The second case is the one where we must determine whether the standard deviation of a population is equal to a pre-established value. Here we limit it to the only presentation of the χ^2 distributional law. On the other hand, we often use Pearson's test of accuracy (χ^2) *and we must be careful not to confuse this with the exponential law that we discuss in the following section.*

1.20. Exponential distribution

Defined based on R^+, this law is a special instance of a gamma law. In reliability, we frequently use exponential law when the failure rate (λ) is constant. This case is highly common in the field of electronics among others. It is also a case that is re-approached in Poisson distribution, but within this a continuous random variable (CRV) is used instead of a discrete random variable (DRV).

This law is of non-negligible importance in mechanical conceptions to the extent that it allows us to estimate the aptitude of a *fully functioning* component once in use. Bell Canada uses it to model the waiting phenomena where the duration of certain services is of exponential nature. In reliability, this law is often used to represent the lifespan of electronic circuits. The mean is often called MTBF. Let us recall that exponential law is a special case of gamma law for $\kappa = 1$. Its principal characteristics are presented below:

Mean, variance, and standard deviation:

$$\mu = \frac{1}{\lambda(\tau)} = E(\tau);\ V = \frac{1}{\lambda^2(\tau)}\ with-inversion\ \tau \succ 0$$

At times, it could seem a little less convincing to consider the mean similar to its standard deviation. Here is a useful small example of this:

Mean E(τ) is written as:

$$E(\tau) = \mu(\tau) = \frac{1}{\lambda(\tau)}\ with\ \tau \succ 0$$

By definition, it is equal to:

$$E(\tau) = \int_{-\infty}^{+\infty} \lambda \times f(\tau) d\tau = \int_{0}^{+\infty} \tau \times \lambda \times \exp^{-\lambda \times \tau} d\tau$$

Part-by-part integration $\int udv = uv - \int vdu$ and a variable change where $[u]$ *represents* $[\tau]$ *and dv represents* $\left[\tau \times \lambda \times Exp\{-\lambda \tau d\tau\}\right]$ allows us to write this as:

$$dv = du, so\ v = \left[\tau \times \lambda \times Exp\{-\lambda \times \tau d\tau\} = -Exp\{-\lambda \tau\}\right]$$

$$E(\tau) = \int_{0}^{+\infty} \tau \times \lambda \times Exp^{-\lambda \times \tau} d\tau = -\tau \times Exp^{-\lambda \times \tau}\Big|_0^{\infty} - \int_{0}^{+\infty} \left(-Exp^{-\lambda \times \tau}\right) d\tau =$$

$$= 0 - \frac{Exp^{-\lambda \times \tau}}{\tau}\Big|_0^{\infty} = \left(\frac{1}{\tau}\right) \text{ or } \mu = E(\tau) = \left(\frac{1}{\tau}\right) \to QED$$

We can also demonstrate that: $\sigma_x^2 = \left(E_x^2 - \mu_x^2\right) = {}^{1}\!/_{\lambda_x^2}$ *with* $\tau \succ 0$. *Why it was necessary to demonstrate?*

In the case of the use of Bell Canada's waiting lists, $\mu = (1/\tau)$ represents a certain mean rate of waiting time. For example, we give $\lambda = 0.5 \to \lambda = \lambda(\tau) = 0.5$, and we have $\theta = 1/\lambda(\tau) = 2$ and standard deviation $\sigma = \theta\ (MTBF) = 1/\lambda = 2$.

The variance will be: $\sigma^2 = \theta^2 = 1/\lambda^2 = 4$ and the mean (MTBF) = $\mu = \theta = 2$.

Failure rate: $\lambda = \lambda(\tau)$ *when* $\tau \geq 0$

Reliability:

$$R(\tau) = Exp\{-\lambda \times \tau\} = Exp\left\{-\frac{\tau}{MTBF}\right\} \text{ with } \tau \geq 0 \quad [1.95]$$

Probability density:

$$f(\tau) = \lambda \times Exp\{-\lambda \times \tau\} = \lambda \times R(\tau) \text{ with } \tau \geq 0 \quad [1.96]$$

We can easily verify that $\int_0^\infty \tau \times Exp\{-\lambda \times \tau\} d\tau = 1$

f(τ) =	g(τ) =	h(τ) =
2	1	0.5
1.637	0.905	0.476
1.341	0.819	0.452
1.098	0.741	0.43
0.899	0.67	0.409
0.736	0.607	0.389
0.602	0.549	0.37
0.493	0.497	0.352
0.404	0.449	0.335
...	...	...

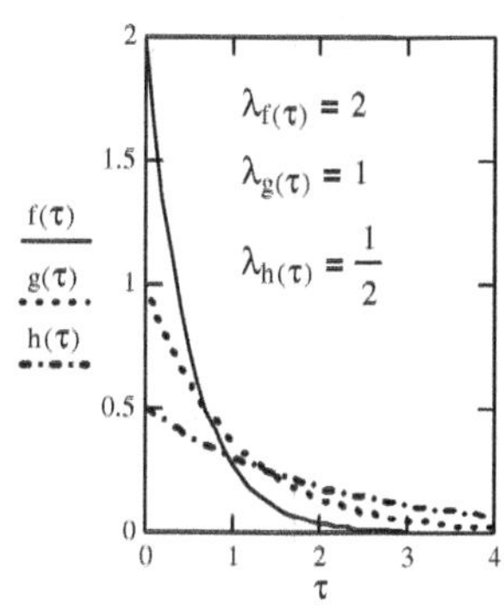

Figure 1.40. *Probability density of exponential law*

The CDF, function of failure or the distribution function is written below:

$$F(\tau) = 1 - Exp\{-\lambda \times \tau\} = 1 - R(\tau) \text{ with } \tau \geq 0 \quad [1.97]$$

$F(\tau) = \int_0^\tau f(u)du = \int_0^\tau \lambda \times Exp\{-\lambda \times \tau\} d\tau$ can be verified more accurately than $P\{T \succ \tau\} = 1 - P\{T \leq \tau\} = Exp\{-\lambda \times \tau\}$.

$dexp(\tau, \lambda_1) =$	$pexp(\tau, \lambda_1) =$
2	0
1.637	0.181
1.341	0.33
1.098	0.451
0.899	0.551
0.736	0.632
0.602	0.699
0.493	0.753
0.404	0.798
0.331	0.835
...	...

Inverse cumulative probability distribution

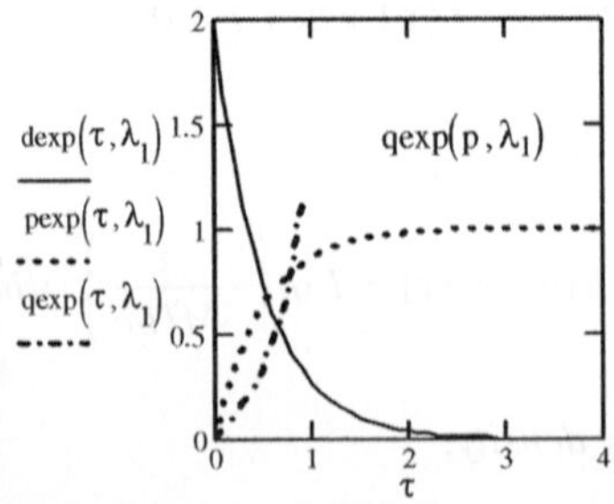

etc. ...

Figure 1.41. *Probability density of inverse exponential law*

Reliability and failure rates:

$R(\tau, \lambda) =$

	0
0	1.000
1	0.951
2	0.905
3	0.861
4	0.819
5	0.779
6	0.741
7	0.705
8	...

etc. ...

Reliability $R(\tau, \lambda) = e^{-\lambda \cdot \tau}$

Failure rate $Z(\tau) = \lambda$

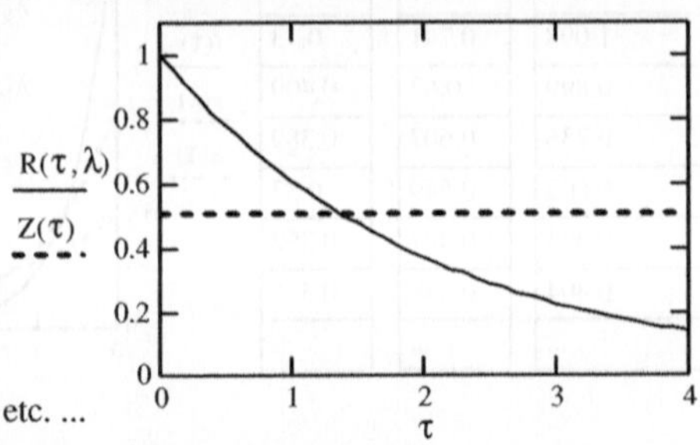

Figure 1.42. *Reliability and failure rates of exponential law*

Hazard values allowing exponential distribution

$$r\exp(m, \lambda_1) = \{\ 0.254, 0.897, 0.398, 1.432, 0.122, 0.327, 0.066\ \}$$

1.20.1. *Example of applying mechanics to component lifespan*

We have machined (*or fabricated*) the following component (see Figure 1.43) according to the standard geometric tolerance.

The design, which we will focus on, will be treated according to resistance calculations and a sound choice of material standards **4010** (molybdenum steel according to SAE nomenclature). We predicted that this component would last **20,000** hours on average if the specifications are correctly observed. It is clear that the distribution of lifespan (τ) of this component follows an exponential distribution, or rather a Weibull distribution with two parameters, where $\beta = 3.7$.

1) Calculate the density and the function of the CDF of probability inherent to the service conditions described above.

2) Estimate the following case studies: P ($\tau \leq 250$ h), P($50 \leq \tau \leq 250$ h), and P($\tau > \mu$).

3) Estimate and comment on the calculation of median lifespan.

4) What could we do if we want to have a lifespan that exceeded the expected duration?

Hypothesis: Having been given the average μ = 20,000 hours, we postulate a mean which is equivalent to $\lambda(\tau)$ thus representing the mean rate of the component's working order.

Demonstration:

1a) The mean of the (τ) is given, so, let us postulate the expression of this mean according to the mean rate of working order, which is presented below:

$$E(\tau) = \lambda(\tau) = \left(\frac{1}{\tau}\right) = \left(\frac{1}{1000}\right) \; hour$$

1b) The PDF of the exponential model **f(τ)** is:

$$f(\tau) = \lambda(\tau) \times Exp^{-\lambda(\tau) \times \tau}, \; so\, f(\tau) = 3.679 \times 10^{-4} \; hours \; of \; service$$

1c) The distribution function of probability of the exponential model $F(\tau)$ is

$$F(\tau) = \left\{1 - \exp^{-\lambda(\tau) \times \tau}\right\} \; so \;\; F(\tau) = 0.632121 \; hours \; of \; service$$

2a) The probability of the exponential model, when τ_1 is $\leq$1001 hours is:

$$P\{\tau_1 \leq 1001\} = F\{1001\} \;\; \text{so} \;\; F(\tau_1) = \left(1 - Exp^{-\lambda(\tau_1) \times \tau_1}\right) = 0.632488$$

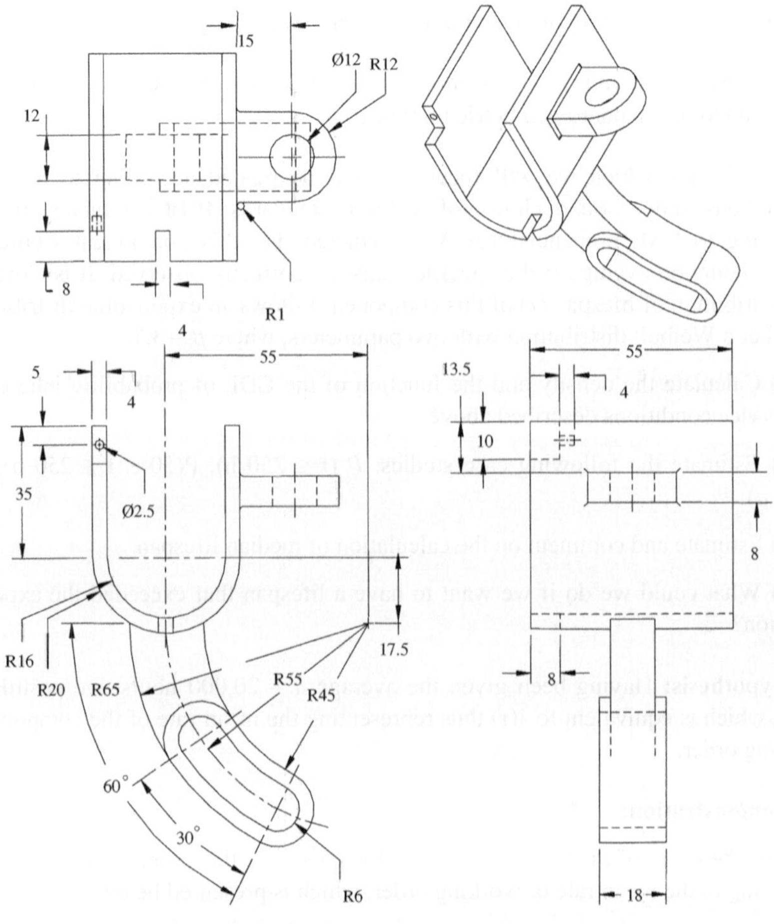

Figure 1.43. *Drawing to show the definition of a real part with SAE4102 standards in inches*

Let us *clarify* that when the part has been working correctly (without failing) for **1001** hours, the probability that it will survive in these conditions is 0.632488, i.e. hardly (63.2488%). Let us call it 63%, that is of all the components, **1001** will still be in working order.

2b) The probability of the exponential model for τ falls in the range $1001 \leq \tau_2 \leq 7001$.

$$\mu_2 = 1000 \text{ } hence \text{ } \lambda_2 = \left(\frac{1}{\mu_2}\right) = 10^{-3}$$

and the probability will be:

$$P\{1001 \le \tau_2 \le 7001\} = \int_{1001}^{7001} \lambda(\tau_2) \times \exp^{-\lambda(\tau_2)\times\tau_2}\, d\tau_2 = 0.366601$$

$P(\tau 2) =$
$2.497 \cdot 10^{-3}$
$3.245 \cdot 10^{-3}$
$3.992 \cdot 10^{-3}$
$4.739 \cdot 10^{-3}$
$5.485 \cdot 10^{-3}$
$6.231 \cdot 10^{-3}$
$6.976 \cdot 10^{-3}$
$7.72 \cdot 10^{-3}$
$8.464 \cdot 10^{-3}$
$9.207 \cdot 10^{-3}$
$9.95 \cdot 10^{-3}$
0.011
0.011
0.012

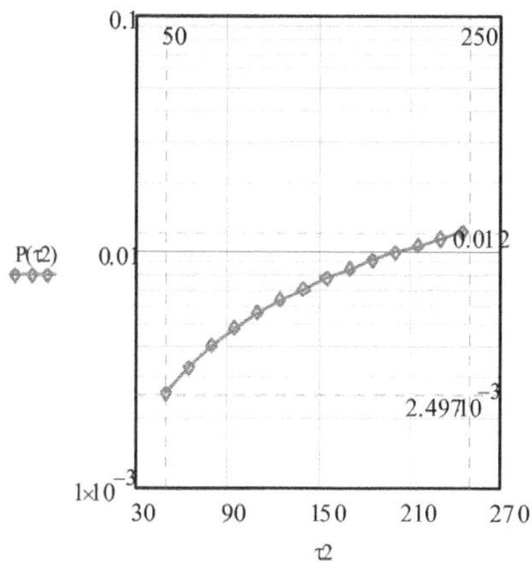

Figure 1.44. *Distribution of probability for τ between 30 and 250*

2c) $P\{\tau_3 \succ \lambda_2\} = 1 - P\{\tau_3 \prec \lambda_2\}$ *because* $1 - F\{\lambda_2\} = 1 - F\{7001\} = 0.367879$

As (100 – 36.6601) = 63.34, let us say that (100 – 36.7879) = 63.212% of the components will have a longevity inferior to the mean.

3c) The median is expressed by the connection shown below:

On one hand $F(med) = \frac{1}{2}$; *and on the other hand*

$$F(med) = \int_{-\infty}^{med} f(\tau_2)\, d\tau_2 = 1 - \exp^{-\lambda(med)}$$

For $F(med) = \frac{1}{2}$, *we also have* $F(med) = 1\text{-}\left\{exp^{-(\lambda_2 \times med)}\right\}$ *which*

can be deduced from this $med = \left[\log\left(\frac{1}{2}\right) \Big/ \log\left(exp^{-(\lambda_2)}\right)\right] = 693.147 \cong 693$ *hours*

COMMENT.– As the median is located *693 hours* on this side of the mean duration of working order (1000 hours), around *30%* of components will have a longevity of less than to 693 hours of working order before failure. To understand whether the law, employed to approach a given physical phenomenon, it is a good idea to proceed to adjustment or adequation. The aim is to find out whether the distribution employed is well adapted to the phenomenon or whether it is only an "insignificant mathematical manipulation". We will return to this in Chapter 3 to look at the adequation theory of testing to understand a few application examples taken from our own dimensional metrology laboratories of quality and industrial automation control.

1.21. Double exponential distribution (Laplace)

The common statistical characteristics of double exponential distribution are:

– Domain $\tau \in \{-\infty, +\infty\}$

– Mean μ, centered median (μ), and mode μ

– Variance $\sigma = \beta\sqrt{2}$, standard deviation $\sigma = 2\beta^2$

– Coefficient of variation $\frac{\beta}{\mu}\sqrt{2}$, skewness = 0, kurtosis = 6

1.21.1. *Estimation of the parameters*

The estimators of *maximum likelihood* of localization and scale parameters of double exponential distribution (also called the Laplace distribution) are:

$$\hat{\mu} = \hat{\tau}\left(\hat{\tau} \text{ is the median of the sample}\right) \text{ and } \hat{\beta} = \left(\sum_{i=1}^{n} \tau_i - \hat{\tau} \Big/ n\right) \quad [1.98]$$

1.21.2. *Probability density function*

The general formula for the probability density function is:

$$f(\tau) = \frac{Exp\left\{-\left|\frac{\tau - \mu}{\beta}\right|\right\}}{2\beta} \quad [1.99]$$

where μ is a localization parameter and β is a scale parameter. For N $[\mu, \beta]$ = N [0, 1], we have the standard double exponential distribution, hence the connection is: $f(\tau) = Exp\{-|\tau|\}/2$

$f(\tau, \mu 0, \beta 0) =$
$2.777 \cdot 10^{-3}$
$2.92 \cdot 10^{-3}$
$3.069 \cdot 10^{-3}$
$3.227 \cdot 10^{-3}$
$3.392 \cdot 10^{-3}$
$3.566 \cdot 10^{-3}$
$3.749 \cdot 10^{-3}$
$3.941 \cdot 10^{-3}$
$4.143 \cdot 10^{-3}$
...

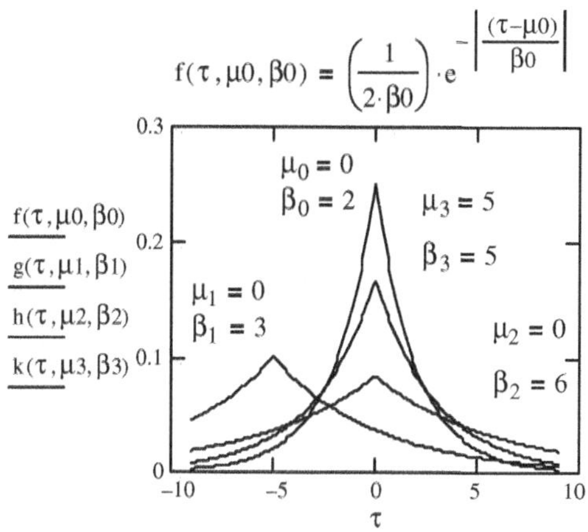

Figure 1.45. *Graph showing the distribution functions of the Laplace distribution*

1.21.3. *Cumulated distribution probability function*

The distribution function of cumulative probability F(τ) is written as:

$$\left\{ \begin{array}{l} F(\tau) = Exp\{-(\tau/2)\} \ for \ \tau \prec 0 \ and \\ F(\tau) = 1 - \{Exp(-\tau)/2\} \ for \ \tau \geq 0 \end{array} \right\} \quad [1.100]$$

The cumulative distribution function F(τ) and its opposite are shown in the following figure:

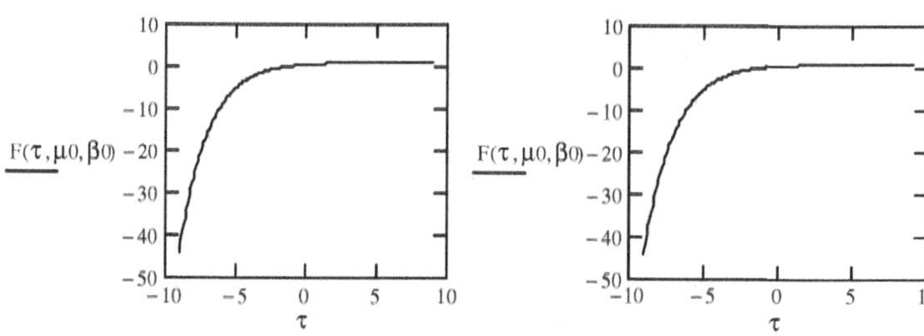

Figure 1.46. *Cumulative distribution function and its opposite, from the Laplace distribution*

Percentile function ***P*(*p*)** of double exponential distribution is:

$$P(p)=\left\{Log(2p)\,for\,p\leq\frac{1}{2}\;\;and\text{ - }Log\{2(1-p)\}\,for\,p\succ\frac{1}{2}\right\}\qquad[1.101]$$

The *risk (chance)* function ***r*(*τ*)** of double exponential distribution (Laplace) is:

$$r(\tau)=\left\{\left(\frac{Exp(\tau)}{2-Exp(\tau)}\right)for\;\tau\prec 0\;and\;1\,for\;\tau\geq 0\right\}\qquad[1.102]$$

Cumulative risk function (chance) ***rc*(*τ*)** of double exponential distribution is written as:

$$r_c(\tau)=\left\{\text{-}Log\left(1-\frac{E\,xp^{\tau}}{2}\right)for\;\tau\prec 0\;and\;\tau+Log(2)\;for\;\tau\geq 0\right\}\qquad[1.103]$$

The *inverse* ***Z*(*P*)** survival function of double exponential distribution (or Laplace distribution) is written as:

$$Z(P)=\left\{Log\{2(1-p)\}\;for\;p\leq\frac{1}{2}\;and-Log(2p)\;for\;p\succ\frac{1}{2}\right\}\qquad[1.104]$$

1.22. Bernoulli distribution

This is also called Bernouilli's sequence (or sometimes *phenomenon*), a series of results of an experiment E that verifies the hypotheses presented below:

a) The experiment can only have two possible results, i.e. occurrence and non-occurrence from a test. The respective probabilities of these results being P and q with q = (1 – P).

b) We also think that the P and q probabilities are constants and that the events (tests) are statistically independent.

c) The tests are independent.

It is a *concrete case study*. Sometimes, for reasons of enonomy of time and money, we can find ourselves drawing from a range of large-scale production, if a component respects the manufactured tolerances (IT = 0.01 inch) without passing by the capabilities. p indicates the probability (identical in each test) that this component will be outside tolerance and q indicates the non-realization of the same event (component accepted in terms of fabrication) at each test.

$t := -5..5 \qquad r(t) = \left(\dfrac{e^{-t}}{2-e^{-t}}\right)$ For $t < 0$

r(τ) =
−1
−1
−1
−1
−1
−1
−1
−1
−1.001
−1.001
−1.001
−1.001
...

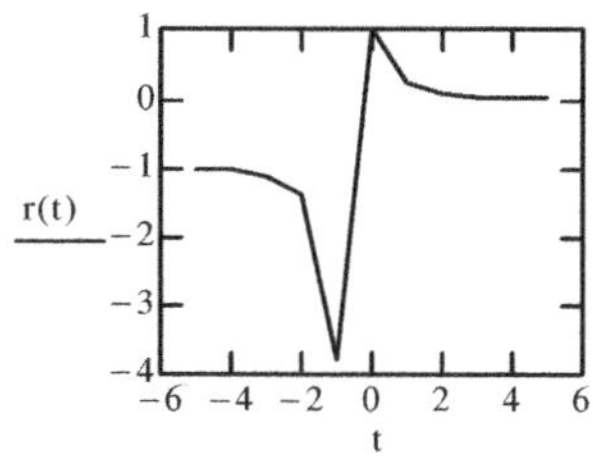

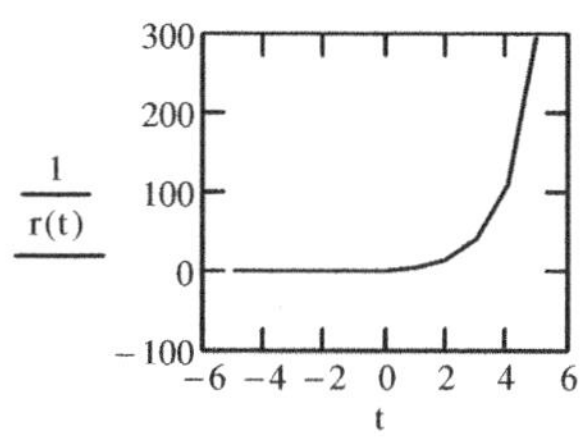

$\frac{1}{r(t)}$ =
−0.987
−0.963
−0.9
−0.729
−0.264
1
4.437
13.778
39.171
108.196
295.826

etc ..

Figure 1.47. *Risk function and its opposite, Laplace distribution*

In reality, we already know that there are only two possible results, whose results are, *de facto*, incompatible (accepted or not, i.e. acceptance of tolerance or rejected tolerance). For this range of manufacturing, the probability is 0.056; we then postulate $(p + q) = 1$.

We have proceeded with $n = 7$ manufactured parts in a large series. Each draw involves the probability p that a part will be outside tolerance (rejected).

As $\tau = 0$, we evaluate the probability that there will be no parts outside of the tolerance range (all of the accepted ITs) in a sample of seven. For n independent events: $\alpha_1, \alpha_2, \alpha_3, \ldots, \alpha_n$, we know that:

$$P\{\alpha_1 \cap \alpha_2 \cap \alpha_3 \cap \ldots_2 \cap \alpha_n\} = P(\alpha_1) \times P(\alpha_1) \times P(\alpha_2) \times P(\alpha_3) \times \ldots \times P(\alpha_n)$$

The probability of all of the IT tolerances succeeding, i.e. 0 defectiveness, out of $\boldsymbol{n} = 7$ when $\mathbf{p} = 0.056$ and $\mathbf{q} = (1 - 0.056) = 0.944$, is written as:

$$P\{\tau = 0\} = (0.944)\times(0.944)\times(0.944)\times(0.944)\times(0.944)\times(0.944)\times(0.944) = (0.944)^7 = 0.668$$

Similarly, even to obtain seven parts outside tolerance for $\tau = 7$, we have:

$$P\{\tau=7\}=(0.056).(0.056).(0.056).(0.056).(0.056).(0.056).(0.056)=(0.056)^7=1.727.10^{-9}$$

Let us calculate the probability of rating a part in a sample of seven. This part could occur anywhere within the seven parts of the sample. For example, the first six parts have succeeded and the seventh part will be eliminated. The problem is postulated as:

$$P\{\tau=1\ out\ of\ 7\}=(0.944)\times(0.944)\times(0.944)\times(0.944)\times(0.944)\times(0.944)\times(0.056)=(0.944)^6(0.056)$$

This rejected part could appear in the third phase of manufacturing (test):

$$P\{\tau=1\ out\ of\ 7\}=(0.944).(0.944).(0.056).(0.944).(0.944).(0.944).(0.944)=(0.056).(0.944)^6$$

and so on.

In the case where these events are independent (not the same manufacturer, not the same machine, etc.), the probability of having one part rejected within a sample of 7 is:

$$P\{\tau=1\ out\ of\ 7\}=7\times(0.056)\times(0.944)^6=0.277$$

The number seven denotes that there are seven different ways of carrying out the manufacturing where we will have one part rejected out of seven. We postulate the following problem:

$$\binom{n}{\tau}=\frac{n!}{\tau!(n-\tau)!}\ ;\ \tau=0,1,2,3,...,until\ n=7$$

Let us deduce, then, that while the number to choose a rejected part among the seven is written as:

$$\binom{7}{1}=\frac{7!}{1!(7-1)!}=\frac{7!}{1!6!}=\frac{7!}{6!}=\frac{7\times6!}{6!}=7$$

For $\tau = 3$, :

$$\binom{7}{3}=\frac{7!}{3!(7-3)!}=\frac{7!}{3!4!}=\frac{7\times6\times5\times4!}{3!4!}=35$$

Each combination has a probability of $(0.056)^2 \times (0.944)^2$ to concretize.

For example: $P\{\tau=3\}=35\times(0.056)^3\times(0.944)^4=4.881\times10^{-3}$.

This continues for the rest of the combinations. Binomial distribution has found a generalized expression to resolve this type of probabilistic problem.

1.23. Binomial distribution

In fact, the Bernoulli schema reproduces itself out of n successive and independent tests. The successes are indicated by (τ) and the failures by ($n - \tau$) without considering the order of appearance. Thus, the probability of having (τ) success on test (n) is:

$$P\{T=\tau\}=\binom{n}{\tau}\times p^{\tau}\times q^{n-\tau}\ ;\ \tau=0,1,2,3,\ldots,until\ n \qquad [1.105]$$

This expression is said to be binomial because it works with two parameters (τ) and (n). Tabulated values exist for this distribution (see Appendices, Volumes 2 and 3, Table A.4). The expression of binomial distribution is written according to [1.100] or as:

$$P\{T=\tau\}=\binom{n}{\tau}\times p^{\tau}\times q^{n-\tau}=b\{\tau;\ n,p\} \qquad [1.106]$$

This distribution is characterized by the following expressions:

Mean: $\mu=n\times p$ and *Variance*: $V=\sigma^2=n\times p\times(1-p)=n\times p\times q$

Standard deviation: $\sigma=\sqrt{n\times p\times(1-p)}=\sqrt{n\times p\times q}$

Estimator: $\hat{\omega}=\left(\frac{\kappa}{n}\right)$, κ is the number of experiments observed out of (n) *tests*

Limits of bilateral confidence of (p), at confidence threshold ($1-\alpha$):

$$For\ n\geq 30\ we\ postulate\ \left(\frac{\kappa}{n}\right)\pm U_{\alpha/2}\times\sqrt{\frac{1}{n}\times\frac{\kappa}{n}\times\left(1-\frac{\kappa}{n}\right)} \qquad [1.107]$$

For $(n) < 30$, we will look at the values provided in the appendix.

Application exercise: During the manufacturing process of embossing, we have embossed many parts in alloys of aluminum, 6061. We have classed 5% of defective parts on a scale of 20 parts.

1) What is the probability of having *two defective* parts?

2) What is the probability of having *more than three* defective parts?

3) What is the probability of having *less than five* defective parts?

4) What is the probability of having *at most four* defective parts?

5) On average, how many parts will be defective out of a sample of *20*?

6) Graphically tracing the binomial distribution law and comparing the results to program with those that come from the standard table reading.

Solution: From the fact that it is a binomial distribution with parameters $n = 50$ and $p = 0.05$ (5%) for reasons previously shown, we postulate the following:

$$b(\tau;20,0.05)=\begin{pmatrix}20\\ \tau\end{pmatrix}\times(0.05)^{\tau}\times(0.95)^{(20-\tau)}\ \textit{with}\ \tau=0,1,2,3,...,20$$

1) The probability of having *two defective parts* is:

$$P\{\tau=2\}=b(2;20,0.05)=\begin{pmatrix}20\\ 2\end{pmatrix}\times(0.05)^{2}\times(0.95)^{(20-2)}=0.188\ 7$$

As the table of binomial distribution (see Appendices) gives ***b* (2, 20, 0.05) = 0.1887**.

2) The probability of having *more than 3* defective parts is:

$$P\{\tau\succ 3\}=b(4;20,0.05)+b(5;20,0.05)+b(6;20,0.05)+...+b(20;20,0.05)=0.016$$

Based on the data given in the binomial distribution table, we now write:

$$=\sum_{\tau=4}^{20}\begin{bmatrix}20\\ \tau\end{bmatrix}\times(0.05)^{\tau}\times(0.95)^{(20-\tau)}=0.0133+0.0022+0.0003+0+0+0+$$

$$+0+0+0+0+0+0++0+0+0+0+0=0.016$$

3) The probability of having *less than five* defective parts is:

$$\left\{\begin{array}{l} P\{\tau \prec 5\} = \sum_{\tau=0}^{\tau \prec 5} \begin{bmatrix} 20 \\ \tau \end{bmatrix} .b(\tau;20,0.05) = b(0;20,0.05) + b(1;20,0.05) + \\ b(2;20,0.05) + ... + b(5;20,0.05) = \sum_{\tau=4}^{4} \begin{bmatrix} 20 \\ \tau \end{bmatrix} .(0.05)^{\tau}.(0.95)^{(20-\tau)} \end{array}\right\}$$

So, $P\{\tau \prec 5\} = 0.3585 + 0.3774 + 0.1887 + 0.0596 + 0.0133 = 0.997\ 0$

Taking into consideration complementary probability, we now write the following:

$$P\{\tau \prec 5\} = 1 - P\{\tau \geq 5\} = 1 - b(5;20,0.05) - b(7;20,0.05) - b(8;20,0.05) -$$

$$-.... - b(5;20,0.05) = 1 - 0.0003 = 0.997\ 0 \textit{ What it was necessary to demonstrate}$$

4) The probability of having *at most three* defective parts is:

$$P\{\tau \leq 3\} = \sum_{\tau=0}^{3} b(\tau;20,0.05) = b(0;20,0.05) + b(1;20,0.05) + b(2;20,0.05) +$$

$$+ b(3;20,0.05) = 0.3585 + 0.3774 + 0.1887 + 0.0596 + 0.0133 = 0.984\ 0$$

5) On average, we will have $\mu = (\text{n} \times \text{p})$ badly embossed (defective) parts:

$\mu = 20 \times (0.05) = 1$ badly embossed part only

6) **Graph**: The results come from our own calculations compared to the values in the tables (see Appendix, binomial distribution table) and are conclusive (identical).

In quality control, or during "statistical enquiries", we are particularly interested in a "portion of success rather than success itself". The reasons for this are obvious. We will not explain them in this context; reliability is the thing that concerns us here. Sometimes, we modify the RV (τ) which here represents the number of successes observed in view of obtaining a relative number, which would translate as a part of the success on n tests. To do this, we devise the VA (τ) by (n). Ultimately, the distribution must *follow* a binomial, as the VA remains invariant, i.e. (τ). The layout of this will be:

$$P\left\{\frac{\tau}{n};n,p\right\} = \binom{n}{\tau}.p^{\tau}.q^{n-\tau} = b\left\{\frac{\tau}{n};\ n,p\right\}.\left(\frac{\tau}{n}\right) \text{ when } 0,\ \frac{1}{n},\ \frac{2}{n},...,1 \qquad [1.108]$$

P(τ) :=

	0
0	0.3585
1	0.3774
2	0.1887
3	0.0596
4	0.0133
5	0.0022
6	0.0003
7	0
8	0
9	0
10	0
11	0
12	...

b(τ, n, p) =

	0
0	0.3585
1	0.3774
2	0.1887
3	0.0596
4	0.0133
5	0.0022
6	0.0003
7	0.0000
8	$2.6590 \cdot 10^{-6}$
9	$1.8659 \cdot 10^{-7}$
10	$1.0803 \cdot 10^{-8}$
11	$5.1688 \cdot 10^{-10}$
12	$2.0403 \cdot 10^{-11}$

Tabulated

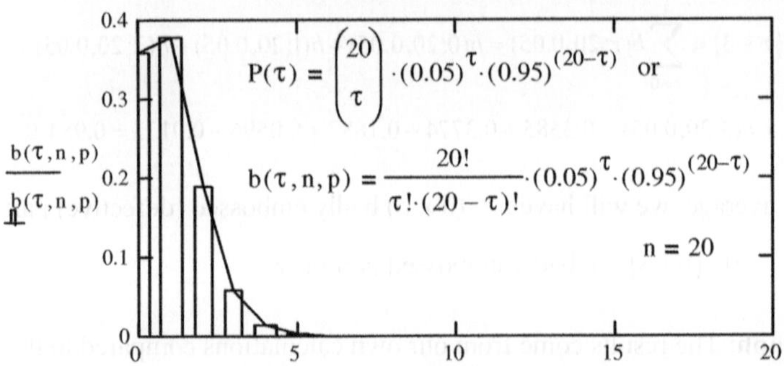

Figure 1.48. *Trace of binomial distribution of badly embossed parts*

So, the *mean* and the *variance* appear in the following way respectively:

$$E\left(\frac{\tau}{n}\right)=\mu\left(\frac{\tau}{n}\right)=\left(\frac{\tau}{n}\right)\times n.p=p \qquad [1.109]$$

$$V\left(\frac{\tau}{n}\right)=\sigma^2\left(\frac{\tau}{n}\right)=\left(\frac{1}{n^2}\right)\times V(\tau)=\left(\frac{1}{n^2}\right)\times n.p.q=\left(\frac{p.q}{n}\right) \qquad [1.110]$$

1.24. Polynomial distribution

It is, in fact, a generalization of binomial distribution in the case of (m) characters $(A_1, A_2, A_3, \ldots, A_m)$ respective of elementary probabilities $(p_1, p_2, p_3, \ldots, p_m)$. We call $(\tau_1, \tau_2, \tau_3, \ldots, \tau_m)$ the (m) VA which represent the respective occurrences of $(A_1, A_2, A_3, \ldots, A_m)$ during (n) tests (identical experiments) formulated below:

$$\sum_{i=1}^{m} \tau_i = n$$

Probability is expressed by the following formula:

$$P\{\tau_1 = \kappa_1, \tau_2 = \kappa_2, \ldots, \tau_m = \kappa_m\} = \left(\frac{n!}{\kappa_1!\ldots\kappa_m!}\right) \times \left(p_1^{\kappa_1}\right) \times \left(p_1^{\kappa_1}\right) \times \ldots \times \left(p_1^{\kappa_m}\right) \qquad [1.111]$$

1.25. Geometrical distribution

Geometrical distribution accepts probability density as its expression of $(q = 0.54;\ \tau = 0, 0.1, \ldots, 6)$ which is:

$$f(\tau) = q \times (1-q)^{\tau} \qquad [1.112]$$

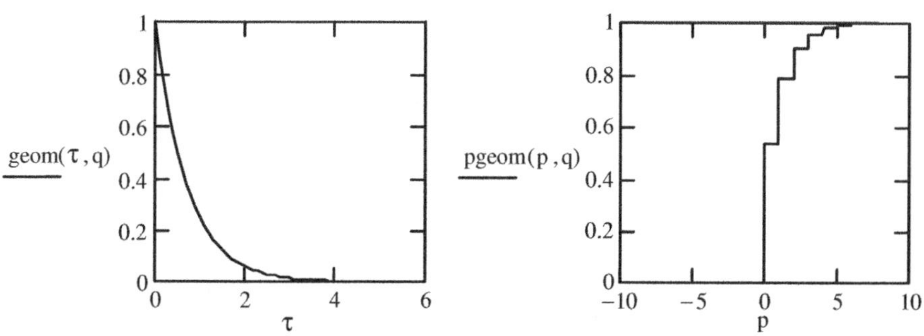

Figure 1.49. *Distribution function and cumulative distribution function*

Vector of random m values which have a geometrical expression is given in the following:

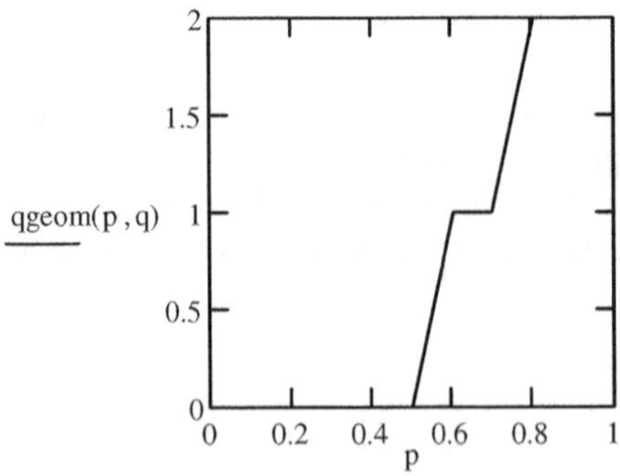

Vector of m random numbers having the geometric distribution

$$\text{rgeom}(m, q) = \begin{pmatrix} 8 \\ 2 \\ 0 \\ 1 \\ 0 \\ 2 \\ 0 \end{pmatrix}$$

For m = 7

Figure 1.50. *Inverse function of cumulative distribution of geometric distribution*

1.25.1. *Hypergeometric distribution (the Pascal distribution) versus binomial distribution*

This distribution with its unusual name is a distribution, which is also named after Pascal (Blaise Pascal, 1652). We use the hypergeometric distribution in cases where the sampling is done with part replacement (failing components, i.e. damaged). Additionally, when the sample size is too small with regard to its population to which p remains invariable, we use hypergeometric distribution to model the distribution of the number of successes (τ). Unlike binomial distribution, the sample is exhaustive and has no remission.

1.25.1.1. *Position of the problem*

We either have a finite population of T components (parts), or a case where (α) represents success and (β) represents failure. T is equal to (α) + (β). Here, (n) is a sample from population T. We aim to determine the probability of having (τ) success and ($n - \tau$) failure in this (n) sample. The first logical idea that comes to mind is to impose a mathematical combination. So, we will first determine the number of *ways* of choosing (τ) success by (α) noted: $\begin{bmatrix} \alpha \\ \tau \end{bmatrix}$ and failure ($n - \tau$) among (β) and noted: $\begin{bmatrix} \alpha \\ n-\tau \end{bmatrix}$.

Together, successes (τ) and failure ($n - \tau$) form a product: $\begin{bmatrix} \alpha \\ \tau \end{bmatrix} \times \begin{bmatrix} \alpha \\ n-\tau \end{bmatrix}$

Moreover, the ways of choosing a sample of size (n) from a population forming N components are written as:

$$\begin{bmatrix} N \\ n \end{bmatrix} = \begin{bmatrix} \alpha+\beta \\ n \end{bmatrix}$$

This reasoning leads us to postulate that:

$$\left\{ H(\tau;n,\alpha,\beta) = \frac{\begin{bmatrix} \alpha \\ \tau \end{bmatrix} = \begin{bmatrix} \beta \\ n-\tau \end{bmatrix}}{\begin{bmatrix} \alpha+\beta \\ n \end{bmatrix}}, \quad \begin{matrix} when\ \tau = 0,1,2,...,n\ if\ n \leq \alpha \\ when\ \tau = 0,1,2,...,\alpha\ if\ n \succ \alpha \end{matrix} \right\} \qquad [1.113]$$

It is clear that the hypergeometric distribution depends on the three factors n, α, and β. The statistical elements which make this distribution distinguishable are the mean and the variance respectively, hence the following formulae from the literature:

$$\mu = \frac{n\times\alpha}{\alpha+\beta};\ \sigma^2 = \frac{n\times\alpha\beta\times(\alpha+\beta-n)}{(\alpha+\beta)^2\times(\alpha+\beta-1)} = \frac{n\times\alpha\beta\times(T-n)}{(\alpha+\beta)^2\times(T-1)} \qquad [1.114]$$

We have deliberately chosen $\left(\frac{T-n}{T-1}\right)$, a factor of correction in (T) which is the finite population. As the Pascal distribution is fastidious to work out, we prefer working with an approximation with the binomial distribution (or the Poisson distribution) which is easier in terms of approach, if well understood $(n/T) \leq 0.1$. We must understand that the sample can exceed 10% of the finite population.

We mean that:

$$p = \left(\frac{\alpha}{\alpha+\beta}\right)\ and\ q = \left(\frac{\beta}{\alpha+\beta}\right)\ knowing\ (p+q) = 1$$

The mean and the variance are:

$$E(\tau) = \left(\frac{n.\alpha}{\alpha+\beta}\right) = n.p\ and\ V(\tau) = \left(\frac{n.\alpha\beta(T-n)}{(\alpha+\beta)^2\times(T-1)}\right) = (n.p.q)\left(\frac{T-n}{T-1}\right) \qquad [1.115]$$

As the mean of the two distributions appears identical, i.e.

$$\left(\frac{n\times\alpha}{\alpha+\beta}\right)=n\times p \text{ if } \left(\frac{n}{T}\right)\leq 0.10$$

the variance of the hypergeometric distribution will lean toward the variance of binomial distribution. Equally, we will note that the more ($\tau = \kappa$) increases, the further apart the gap becomes between hypergeometric distributions, binomial distributions, and Poisson distributions. We will now summarize the point: When T is high, the hypergeometric distribution becomes significantly more similar to a binomial distribution of the same mean.

Practical case study of hypergeometric distribution: Producing **25** splined shafts (from the 1st tree to the 25th) in parallel. In the manufacturing range, there was an error due to a dimensional non-transfer. Inadvertently, the manufacturer substituted (α) three parts with three other parts from another testing area which had neither been used that time, nor in the same setup conditions. Instead of continuing manufacturing, the manufacturer drew out *three parts* at random and tested them using metrology. If one of these parts, *at most*, had shown itself to be outside tolerances, the manufacturer would have had to restart all of the manufacturing ranges.

1) What is the probability that on reaching the 10th part, the manufacturer will no longer have to redo the manufacturing range? Illustrate your response using a graph.

2) What is the probability that on reaching the third range, the manufacturer will have to redo the manufacturing? (accepting that his rating is at the maximum of the material)

Response: We can clearly see that at part 10, the manufacturer has no chance of restarting the manufacturing (see graph). On the other hand, at part eight, the manufacturer has a probability of q = (1 – p) of restarting the manufacturing process, at the maximum of material p = (1 – 0.0022) = **0.998** noticeably, only one chance of restarting manufacturing, etc (see graph).

1.26. Hypergeometric distribution (the Pascal distribution)

Hypergeometric distribution is often used for simultaneous production (i.e. non-ordered and without remission) of *n* elements among *A*. We would consider this production to be equiprobable. This distribution is often approached by binomial distribution with parameters (*n, p*) as soon as *n*/*A* < 10%, i.e. when the sample *n* is 10 times smaller than population *A*.

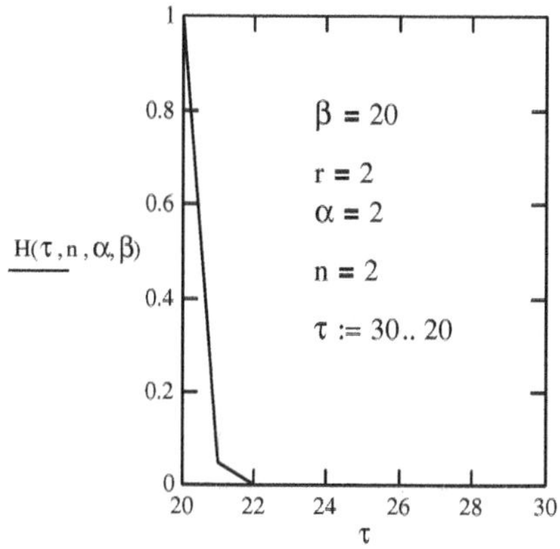

H(τ, n, α, β) =
$9.172 \cdot 10^{-15}$
$2.7516 \cdot 10^{-13}$
$7.9796 \cdot 10^{-12}$
$2.2343 \cdot 10^{-10}$
$6.0326 \cdot 10^{-9}$
$1.5685 \cdot 10^{-7}$
$3.9212 \cdot 10^{-6}$
$9.4109 \cdot 10^{-5}$
0.0022
0.0476
1

Figure 1.51. *Graph showing hypergeometric distribution versus binomial distribution*

Distribution and cumulative distribution for the hypergeometric distribution are shown in Figure 1.52:

phypergeom(m, α, β, n) =
0.823
0.996
1
1
1
1
1
1

dhypergeom(m, α, β, n) =
0.823
0.173
$4.329 \cdot 10^{-3}$
0
0
0
0
0

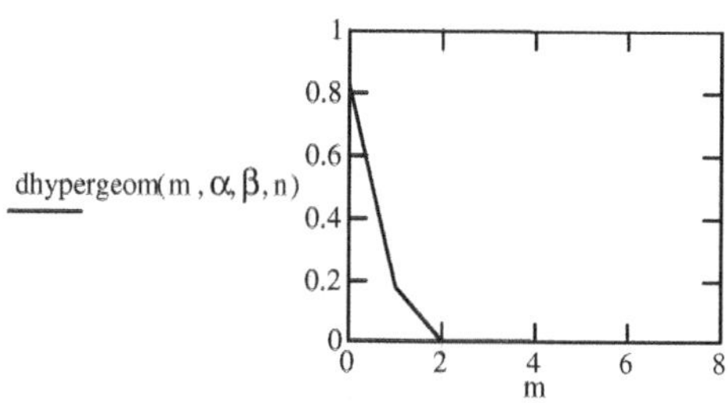

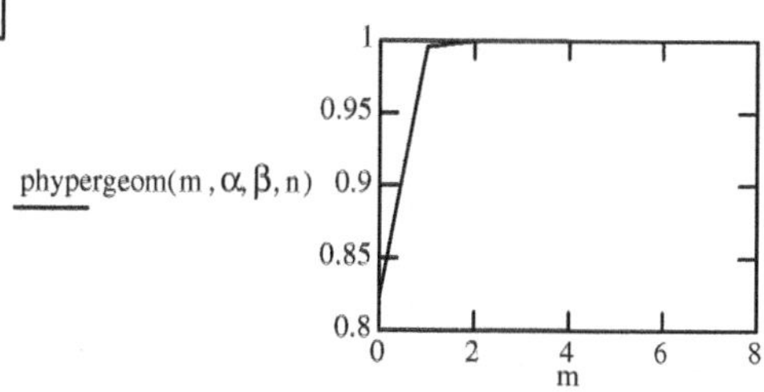

For $\tau := 30..20$

$\beta = 20$ $n = 2$ $r = 2$

$\alpha = 2$

$p := 0, 0.1..1$

Figure 1.52. *(dhyper(.)) and phyper(.)) distribution functions for hypergeometric distribution (Pascal distribution)*

1.27. Poisson distribution

This distribution (by the mathematician Simon Poisson 1781–1840) essentially depends on two parameters: $\kappa > 0$ and $\lambda > 0$. The befalling of a phenomenon such that single events can happen one at a time (non-simultaneity of realizations) and such that the number of events that happen during a period, τ, only depends on the duration of this period. In our area of study, we will consider the Poisson distribution to be, in fact, only an approximation of binomial distribution. When (n) the number of tests increases, probability evaluation will be high.

Probability density; (*nxp*) = Constant

The probability of achieving (τ) success in (n) number of trials (once again, using the Bernoulli schema) when (n) is high and (p) is small is:

$$\lim_{n\to\infty} b(\tau;n,p)=\frac{(n.p)^{\tau}}{\tau!}\cdot Exp\{-n\times p\}=p(\tau;np) \qquad [1.116]$$

The approximation of binomial distribution is satisfactory if ($n.p$) < 5 and p ≤ 0.1. So, when postulating (λ) = ($n.p$), we will formulate the Poisson distribution as:

$$p(\tau;n,p)=p(\tau;\lambda)=\left(\frac{\lambda^{\tau}}{\tau!}\right)\times Exp\{-\lambda\} \qquad [1.117]$$

We notice that the Poisson distribution only depends on one parameter, λ, which represents its variance and which is also its mean. This law is characterized by the following expressions:

– *Mean*: $\mu = n\times p = E(\tau)=\lambda$

– *Variance*: $\sigma^2 = E(\tau-\mu)^2 = np = \lambda$

– *Standard deviation*: $\sigma=\sqrt{\sigma^2}=\sqrt{np}=\sqrt{\lambda}$

– *Estimator*: $\hat{\omega}=\kappa/n$, κ is the number of experiments observed in (n) trials

COMMENT.– Calculated binomial distribution b(τ, n, p) is equivalent to the tabulated values of the same binomial distribution. Poisson P(τ, 2) is equivalent to the values tabulated for the same Poisson distribution when calculated: We notice that the Poisson distribution P(τ) coincides with binomial distribution in the same conditions. Sometimes, we may approximate binomial probabilities using the Poisson distribution. With the aim of showing the excellent coming together of these two distributions, we will show by a practical example how to mathematically and graphically illustrate this concordance. Here is the position of our problem:

$b(\tau, n, p)$ = Binomial	$P(\tau)$ = Poisson
0.12158	0.13534
0.27017	0.27067
0.28518	0.27067
0.19012	0.18045
0.08978	0.09022
0.03192	0.03609
0.00887	0.01203
0.00197	0.00344
0.00036	...
...	

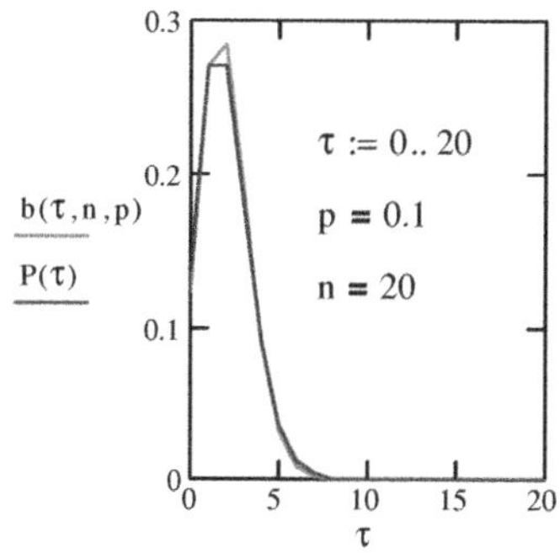

etc..

Figure 1.53. *Graph showing the approximation of binomial distribution by Poisson (or exponential) distribution*

That (λ) = ($n.p$) = 2 has certain variations of (n), up to conditions where p becomes ≤ 0.1 as above is explained in theory. We already know that (λ) = ($n.p$) < 5 and that the condition of study will be satisfied.

1.28. Gamma distribution

This distribution essentially depends on two parameters, $\kappa > 0$ and $\lambda > 0$. It comes closer to the *Erlang* distribution and the Poisson distribution for positive integer κ. Gamma distribution (see Appendices, Volumes 2 and 3, Table A.1) is, in fact, a generalization of exponential distribution, which corresponds to temporal probability distribution separating the appearance of the two given events. This distribution provides probability distribution with the time that passes between the Kth and the $(K + r)$th appearance of the event (trial). We use mechanical equipment subject to use of a model life duration.

The failure rate $Z(\tau)$ follows the value (κ), so we postulate the conditions:

– $\kappa = 1$, so $Z(\tau)$ = constant = λ

– $\kappa > 1$, so $Z(\tau)$ increases with τ from 0 to λ

– $\kappa < 1$, so $Z(\tau)$ decreases τ from $+\infty$ to λ

Therefore, we postulate the following main estimators:

$$Mean\text{: } \mu = \left(\frac{\kappa}{\lambda}\right)\text{; Variance } V = \left(\frac{\kappa}{\lambda^2}\right)\text{; Standard deviation } \sigma = \left(\frac{\sqrt{\kappa}}{\lambda}\right)$$

When (κ) is an integer, we observe a duality in the gamma function with the Poisson function.

Probability density:

$$f(\tau) = \left(\frac{1}{\Gamma(\kappa)}\right) \times \lambda \times \text{Exp}\{-\lambda.\tau\} \times (\lambda.\tau)^{\kappa-1} \quad [1.118]$$

With $\Gamma(\kappa) = \int_0^{\infty} Exp\{-x\} \cdot x^{\kappa-1} dx$ as a type 2 Eulerian function

CDF:

$$F(\tau) = \left(\frac{\lambda^{\kappa}}{\Gamma(\kappa)}\right) \times \int_0^{\tau} u^{\kappa-1} \cdot Exp\{-\lambda.u\} du \quad [1.119]$$

Probability density (another way of expressing it):

$$f(\tau) = \left(\frac{1}{(\kappa-1)!}\right) \times \text{Exp}\{-\lambda.\tau\} \times (\lambda.\tau)^{\kappa-1} \quad [1.120]$$

CDF (another way):

$$F(\tau) = (1 - \text{Exp}\{-\lambda.\tau\}) \times \sum_{\tau=0}^{\kappa=1} \left(\frac{\lambda\tau^i}{i!}\right) \quad [1.121]$$

$$Reliability\text{: } R(\tau) = \text{Exp}\{-\lambda.\tau\} \times \sum_{\tau=0}^{\kappa=1} \frac{\lambda\tau^i}{i!} \quad [1.122]$$

$$Failure\ rate\text{: } Z(\tau) = \lambda \times \frac{\lambda\tau^{\kappa-1}}{(\kappa-1)! \left(\sum_{\tau=0}^{\kappa=1} \frac{(\lambda\tau)^i}{i!}\right)} \quad [1.123]$$

f(τ) =	g(τ) =	h(τ) =
0.09	0.905	1.614
0.129	0.861	1.254
0.164	0.819	1.033
0.195	0.779	0.879
0.222	0.741	0.763
0.247	0.705	0.672
0.268	0.67	0.598
0.287	0.638	0.536
0.303	0.607	0.484
0.317	0.577	0.439
0.329	0.549	0.4
...	...	...

$$\mathrm{Erlang}(\tau,\lambda) := \left(\frac{1}{\Gamma(\kappa)}\right)\cdot\lambda\cdot e^{-\lambda\cdot\tau}\cdot(\lambda\cdot\tau)^{\kappa-1}$$

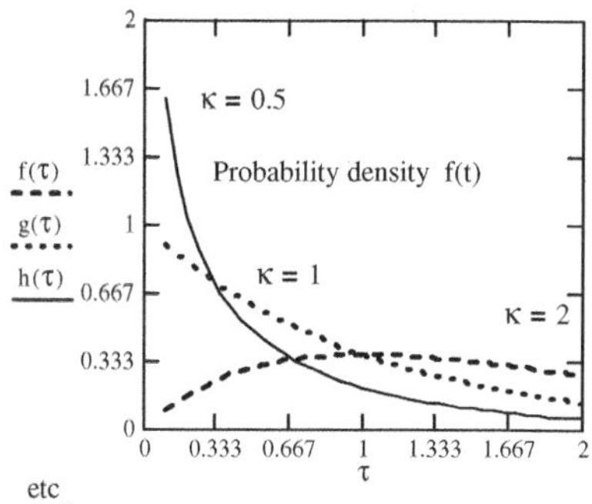

etc..

Figure 1.54. *Probability density of the gamma distribution (λ=1) (Erlang distribution)*

dgamma(τ,λ) =	pgamma(τ,λ) =
0	0
0.861	0.139
0.741	0.259
0.638	0.362
0.549	0.451
0.472	0.528
0.407	0.593
0.35	0.65
0.301	0.699
0.259	0.741
0.223	0.777
0.192	0.808
0.165	0.835
0.142	0.858

$$G(\tau,\lambda) = \frac{\left(\tau^{\lambda-1}\cdot e^{-\tau}\right)}{\Gamma(\lambda)}$$

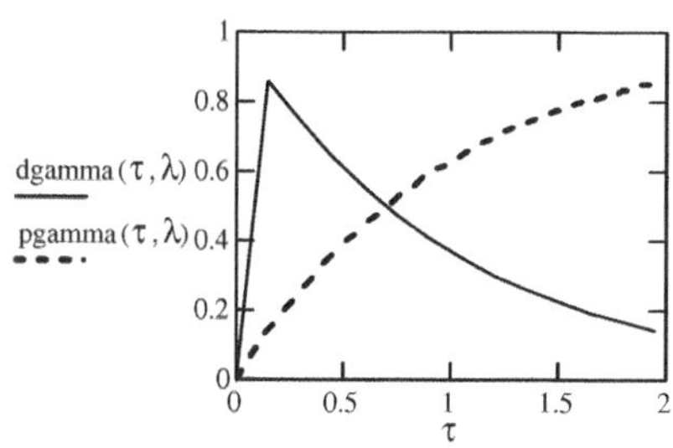

$\tau := 0, 0.15 .. 2$ $\lambda = 1$

Probability density, dgamma(τ,λ)

Cumulative probability distribution, pgamma(τ,λ)

Figure 1.55. *Distribution and cumulative distribution functions and gamma distribution probability*

A connection between the CDF, gamma distributions (also known as the *Erlang* distribution), and the Poisson distribution exists, which is given as:

$$F_{gamma}\{\tau;\kappa;\lambda\} \equiv 1 - F_{Poisson}\{m = \lambda\cdot\tau \ et\ c = \kappa - 1\} \qquad [1.124]$$

qgamma(p, λ) = Inverse cumulative probability distribution for probability p (or 1/F(t, ?)

qgamma(p, λ)
0.105
0.117
0.128
0.139
0.151
0.163
0.174
0.186
0.198
0.211
0.223
0.236
0.248
0.261
...

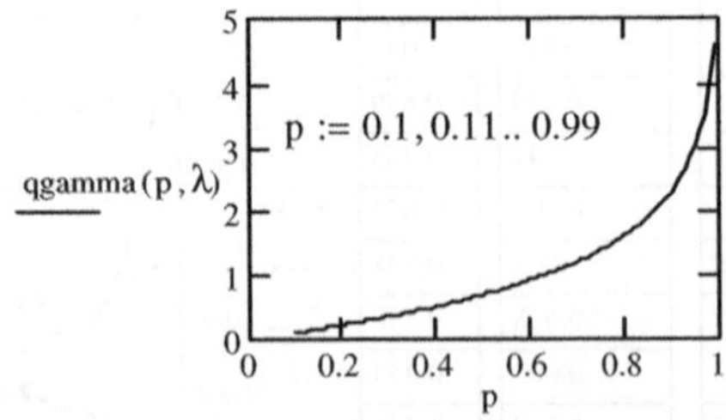

$$m = 7 \qquad \text{rgamma}(m, \lambda) = \begin{pmatrix} 6.67 \\ 1.643 \\ 0.536 \\ 1.049 \\ 0.195 \\ 1.748 \\ 0.342 \end{pmatrix}$$

etc..

Figure 1.56. *Inverse cumulative distribution function of the probability of gamma law and chance production of m=7 of m values hence the law will be distributed according to a gamma distribution*

Reliability R(τ) appears as:

Reliability R(τ) appears as:

R3(τ) =	R4(τ) =	R5(τ) =
1	0.995	0.905
0.999	0.99	0.861
0.999	0.982	0.819
0.998	0.974	0.779
0.996	0.963	0.741
0.994	0.951	0.705
0.992	0.938	0.67
0.989	0.925	0.638
0.986	0.91	0.607
...	...	...

$$R5(\tau) = e^{-\lambda \cdot \tau} \cdot \sum_{i=0}^{\kappa_5 - 1} \frac{(\lambda \cdot \tau)^i}{i!}$$

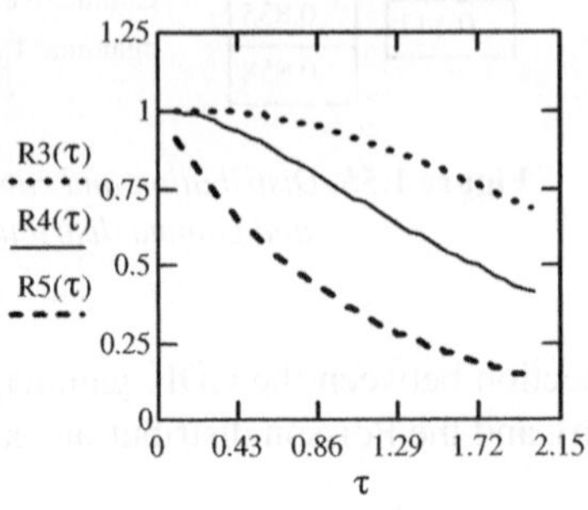

Figure 1.57. *Gamma distribution reliability (Erlang with λ=1)*

1.29. Inverse gamma distribution

Inverse-gamma distribution forms part of the continuous probability distributions with two parameters on the almost straight line of real positives. It is the opposite of a random variable, distributed according to gamma distribution.

Parameters (λ, β): *Shape parameter* $\lambda \succ 0$ *and* scale *parameter* $\beta \succ 0$

Mean: $\left(\frac{\beta}{\lambda-1}\right)$ *for* $\lambda \succ 0$; *mode:* $\left(\frac{\beta}{\lambda+1}\right)$; and variance: $\left(\frac{\beta^2}{(\lambda-1)^2(\lambda-2)}\right)$ *for* $\lambda \succ 2$

Skewness: $\left(\frac{4\sqrt{\lambda-2}}{\lambda-3}\right)$ *for* $\lambda \succ 3$; and *kurtosis* $6\times\left(\frac{5\lambda-11}{(\lambda-3)\cdot(\lambda-4)}\right)$ *for* $\lambda \succ 4$

1.30. Distribution function (inverse gamma distribution probability density)

The distribution function, i.e. probability density of inverse gamma distribution, is defined on the support $\tau > 0$. It is presented as:

$$f(\tau;\lambda,\beta) = \frac{\beta^{\lambda}}{\Gamma(\lambda)}\left(\frac{1}{\tau}\right)^{\lambda+1} \times E\text{xp}\left\{-\left(\frac{\beta}{\tau}\right)\right\} \qquad [1.125]$$

In this case, λ is a shape parameter and β is a parameter of intensity, i.e. the opposite of a scale parameter. The domain is $\tau \in (0;+\infty]$.

The CDF (inverse gamma):

$$F(\tau;\lambda,\beta) = \left\{\Gamma\left(\lambda,\frac{\beta}{\tau}\right)\Big/\Gamma(\lambda)\right\} \qquad [1.126]$$

For: $\lambda_1 = 2$; $\beta_1 = 0.5$; $\lambda_2 = 3$; $\beta_2 = 1.5$; $\lambda_3 = 4$; $\beta_1 = 3$; $\lambda_4 = 1$; $\beta_4 = 1$

1.31. Erlang distribution (characteristic of gamma distribution, Γ)

Continuous function, the Erlang distribution (Agner Krarup Erlang), is essentially used for the coming together of distributions such as exponential and gamma distributions. Moreover, the Erlang distribution is no other distribution but the gamma distribution with integer κ. We particularly use this to make two waiting-line models approach each other, in simultaneous telephony. It is characterized by two parameters (binomial): we must know that κ is a shape parameter, a positive integer, and that λ is a real parameter called intensity. In an alternative parameterization, we consider η as a scale parameter equivalent to $1/\lambda$. When the parameter of form κ is worth 1, the distribution comes directly closer to the exponential distribution, as does binomial distribution.

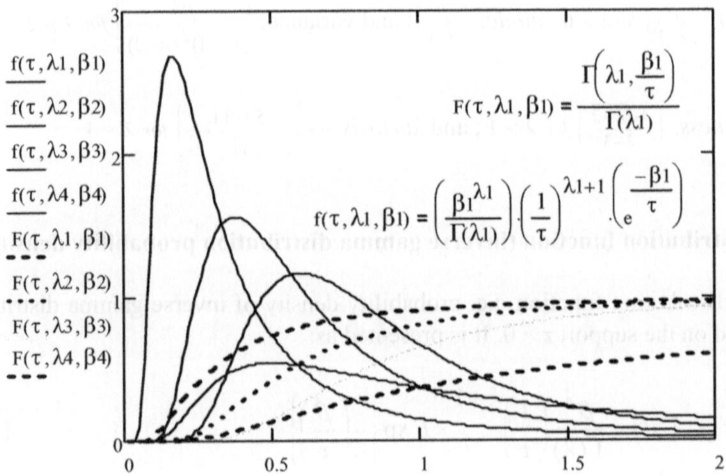

Figure 1.58. *Distribution functions and the cumulative distribution of inverse gamma distribution*

The Erlang distribution is a special case of gamma distribution. In this precise case, κ is an integer (form of distribution). However, we have seen that in gamma distribution, parameter κ is a real positive. In summary, the parameters from the literature which distinguish the Erlang distribution appear in the following way:

– $\kappa \succ 0$ is a shape parameter, positive integer

– $\lambda \succ 0$ is a real positive, qualified by *intensity*

– $\eta = (1/\lambda) \succ 0$ is a real positive integer also called scale parameter

In its general form, especially when $k < 1$, the Erlang distribution represents the period of *youth* and of *useful life* of components. Equally, this distribution is used to approach problems of sequential redundancy and chain failure (offshore structures, series resistances, etc.) The main statistical characteristics of this distribution are the following:

$$Mean = E(\tau) = \frac{\kappa}{\lambda};\ variance = \sigma^2 = \left(\frac{\kappa}{\lambda^2}\right)\ and\ skewness\ A_{skewness} = \left(\frac{2}{\kappa}\right)$$

Probability density (or function of mass):

$$f(\tau;\kappa,\lambda) = \frac{\left(\lambda^{\kappa}\right)\times\left(\tau^{\kappa-1}\right)\times Exp\{-\lambda.\tau\}}{(\kappa-1)!}\ for\ \tau \succ 0, \tau \in [0;\ \infty) \qquad [1.127]$$

An equivalent parameterization would bring in η, a scale parameter, defined as "an inversement proportional to intensity η", i.e. η = 1/λ:

$$f(\tau;\kappa,\eta) = \frac{\left(\tau^{\kappa-1}\right)\times Exp\left\{-\frac{\tau}{\eta}\right\}}{\eta^{\tau}\times(\kappa-1)!}\quad for\ \tau \succ 0 \qquad [1.128]$$

$(\kappa - 1)!$ → lay upon (κ) being a natural number.

CDF:

$$F(\tau;\kappa,\lambda) = \frac{\Gamma\times(\kappa,\lambda\cdot\tau)}{(\kappa-1)!} = 1 - \sum_{i=0}^{\kappa-1} Exp\{-\lambda.\tau\}\frac{(\lambda\cdot\tau)^i}{i!} \qquad [1.129]$$

where Γ(.) is the incomplete gamma function.

While studying of the Poisson distribution, we see that the events, which occur with an average given intensity (occurrence), are modeled here.

Graphs showing the probability density for different κs:

The waiting times between κ occurrences are distributed according to an Erlang distribution. The listing of events in (τ) is taken into account by the Poisson distribution. In the *stochastic processes*, Erlang distribution constitutes a distribution of the sum of (κ_*RV*_*i.i.d.*), i.e. *random and identically, independently distributed variables* according to an exponential distribution.

We can clearly see the graphical and statistical coming together of exponential distribution, gamma distribution, and the Erlang distribution. We will present the estimation of the Erlang distribution in detail in Chapter 3.

$f(\tau) =$
2
1.637
1.341
1.098
0.899
0.736
0.602
0.493
0.404
0.331
0.271
0.222
0.181
0.149
0.122
...

$$f(\tau) = \frac{\left[\left[(\lambda_1)^{\kappa_1}\right]\cdot\left(\tau^{\kappa_1-1}\right)\cdot e^{-\lambda_1\cdot\tau}\right]}{(\kappa_1-1)!}$$

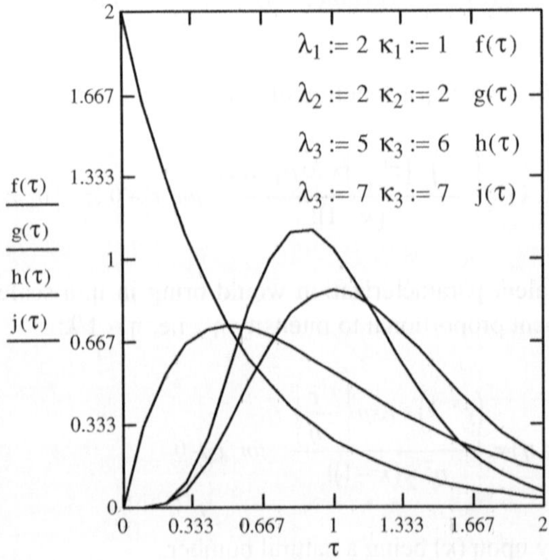

Figure 1.59. *Impression of the density function of Erlang probability*

$F8(\tau) =$	$F7(\tau) =$	$F6(\tau) =$
0	0	0
3.549	0.947	0.035
5.226	2.246	0.283
6.242	3.314	0.752
6.936	4.158	1.317
7.444	4.835	1.883
7.836	5.39	2.412
8.147	5.853	2.893
8.401	6.246	3.326
8.613	6.584	3.717
8.792	6.877	4.072
8.945	7.134	4.396
...	...	...

$$F8(\tau) = \lambda\cdot\frac{(\lambda\cdot\tau)^{\kappa_8-1}}{(\kappa_8-1)!\cdot\sum_{i=0}^{\kappa_8-1}\frac{(\lambda\cdot\tau)^i}{i!}}$$

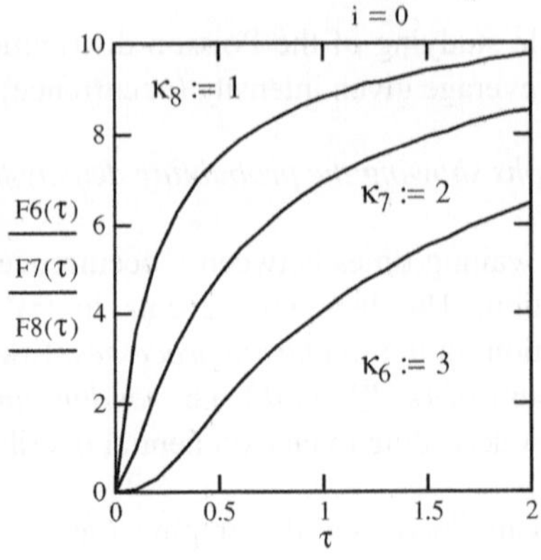

Figure 1.60. *Impression of the cumulative distribution function of the Erlang distribution*

1.32. Logistic distribution

Statistical–mathematical characteristics of logistic distribution: The term logistic distribution comes from its use in logistics due to its CDF which is called logistic function. Below are its statistical characteristics:

$$Support: \tau \in]-\infty,+\infty\ [\ ;\ \mu = E(T) = Mode\ =\ median\ and\ Variance\ =\ VA = \frac{(\pi\lambda)^2}{3},$$

$$\mu = 0\ and\ \lambda = 1 \rightarrow VA = \frac{\pi^2}{3}, Skewness = 0\ and\ normalized\ kurtosis = \left(\frac{6}{5}\right)$$

Simple logistic law with parameters (μ, λ) > 0 presents probability distribution as:

$$f(\tau) = \frac{Exp\left(-\frac{\tau-\mu}{\lambda}\right)}{\lambda \times \left(1 + Exp\left(-\frac{\tau-\mu}{\lambda}\right)\right)^2} \qquad [1.130]$$

Its CDF appears as:

$$F(\tau) = \left[1 + Exp\left(-\frac{\tau-\mu}{\lambda}\right)\right]^{-1} \qquad [1.131]$$

The *standard* function is a logistic distribution with parameters [0 and 1]. In fact, its CDF represents a sigmoid function with the expression:

$$F(\tau) = \left[1 + Exp\{-\tau\}\right]^{-1} \qquad [1.132]$$

The *probability density function* f(τ) (of mass) is traced in this way for different (μ, λ) > 0.

Below is the CDF F(τ), traced for the same parameters:

$(\mu, \lambda) > 0$

Inverse CDF of logistic distribution: For p = [0 to 10] and m = 7.

dlogis(τ, λ1, s1) =

$1.67\cdot10^{-5}$
$1.846\cdot10^{-5}$
$2.04\cdot10^{-5}$
$2.254\cdot10^{-5}$
$2.491\cdot10^{-5}$
$2.753\cdot10^{-5}$
$3.043\cdot10^{-5}$
$3.363\cdot10^{-5}$
$3.717\cdot10^{-5}$
$4.108\cdot10^{-5}$
$4.54\cdot10^{-5}$
$5.017\cdot10^{-5}$
...

$$f(\tau,\lambda,s) = \frac{e^{-\left(\frac{\tau-\lambda}{s}\right)}}{s\cdot\left[\lambda + e^{-\left(\frac{\tau-\lambda}{s}\right)}\right]^2}$$

s1 = 1 λ1 = 1

s2 = 2.5 λ2 = 1.5

s3 = 3.2 λ3 = 2.5

s4 = 1.5 λ4 = 3.5

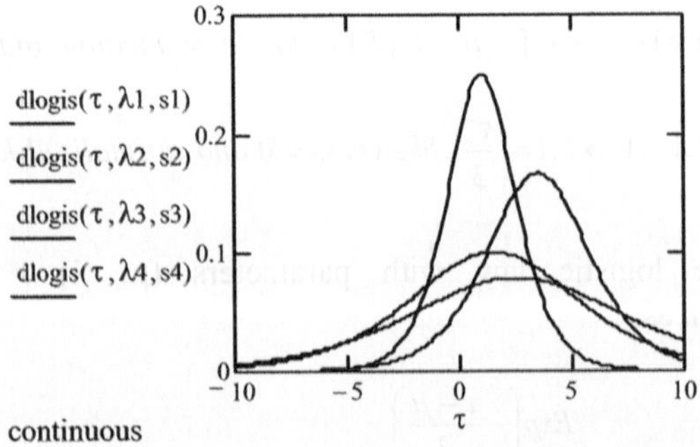

continuous

Figure 1.61. *Graphs showing the distribution function of logistic distribution for τ = [0 to 2], λ_i = [1, 1.5, 2.5, 3.5], and s_i = [2, 2.5, 3.2, 1.5]*

plogis(τ, λ1, s1) =

$1.67\cdot10^{-5}$
$1.846\cdot10^{-5}$
$2.04\cdot10^{-5}$
$2.254\cdot10^{-5}$
$2.492\cdot10^{-5}$
$2.754\cdot10^{-5}$
$3.043\cdot10^{-5}$
$3.363\cdot10^{-5}$
$3.717\cdot10^{-5}$
...

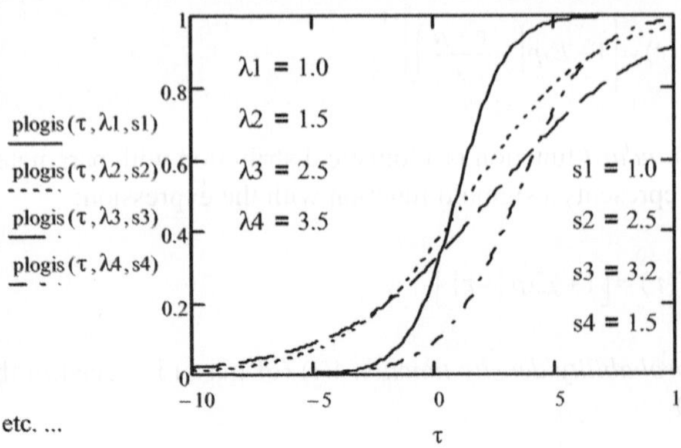

etc. ...

Figure 1.62. *Graph showing the logistic cumulative distribution for τ = [0 to 2], λ_i = [1, 1.5, 2.5, 3.5], and s_i = [2, 2.5, 3.2, 1.5]*

Random numbers (m) allowing a logistic distribution trace (m = 7):

$\mathrm{rlogis}(m,\lambda_1,s_1) = \{-0.014;\ 2.659;\ 0.493;\ 1.741;\ -3.722;\ 0.035;\ 1.355\}$ *etc...*

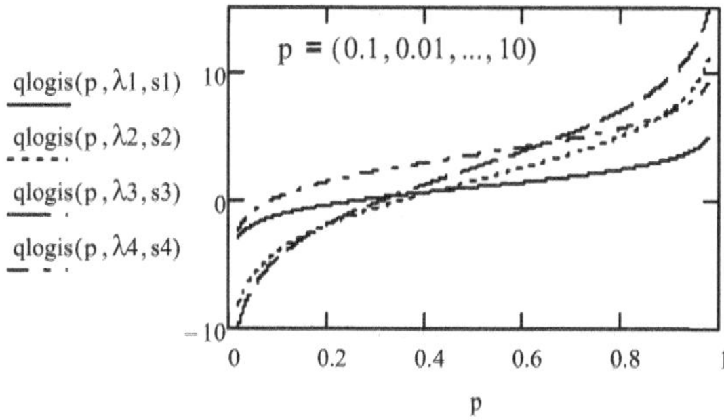

Figure 1.63. *Graph showing the inverse cumulative distribution function of logistic distribution for τ = [0 to 2] and λ_i = [1, 1.5, 2.5, 3.5] and s_i = [2, 2.5, 3.2, 1.5]*

1.33. Log-logistic distribution

1.33.1. *Mathematical–statistical characteristics of log-logistic distribution*

$$Support\ \tau \in [0,+\infty);\ E(T) = \left(\frac{\alpha\pi}{\beta}\right) \Big/ Sin\left(\frac{\pi}{\beta}\right)\ for\ \beta \succ 1\ otherwise\ undefined$$

$$Scale\ parameter: \alpha \succ 0\ and\ shape\ parameter: \beta \succ 0;\ Median = \alpha$$

$$Mode = \alpha \times \left(\frac{\beta-1}{\beta+1}\right)^{\frac{1}{\beta}}\ for\ \beta \succ 1\ otherwise\ undefined$$

For $x > 0$, $\alpha > 0$, and $\beta > 0$, the log-logistic distribution F(τ) is:

$$f(\tau;\alpha,\beta) = \frac{\left(\frac{\beta}{\alpha}\right) \times \left(\frac{\tau}{\alpha}\right)^{-\beta-1}}{\left(1+\left(\frac{\tau}{\alpha}\right)^{-\beta}\right)^2} \qquad [1.133]$$

For $x > 0$, $\alpha > 0$, and $\beta > 0$, the log-logistic CDF F(τ) is:

$$F(\tau;\alpha,\beta) = \frac{1}{1+(\tau/\alpha)^{-\beta}} = \frac{(\tau/\alpha)^{\beta}}{1+(\tau/\alpha)^{\beta}} = \left(\frac{\tau^{\beta}}{\alpha^{\beta}+\tau^{\beta}}\right) \qquad [1.134]$$

The distribution is unimodal when $\beta > 1$ and its dispersion decreases when β increases.

1.33.2. *Moment properties*

The moment of the *k*th order exists only when $\kappa < \beta$, and presents itself as:

$$E\left(T^{\kappa}\right)=\alpha^{\kappa} B\left(1-\frac{\kappa}{\beta};\ 1+\frac{\kappa}{\beta}\right)=\alpha^{\kappa}\times\frac{\kappa\pi}{\beta}\Big/\sin\left(\frac{\kappa\pi}{\beta}\right) \qquad [1.135]$$

where B(.) is the beta function which considers the statistical–mathematical characteristics of logistic distribution.

When $\theta = \pi/\beta$ postulating $b = \pi/\beta$, the mean takes this form:

$$\mathrm{E}(T)=\frac{\alpha\theta}{Sin(\theta)}\ with\ \beta>1\ and\ \mathrm{VAR(T)}=2\theta\Bigg/\left(Sin(2\theta)-\frac{\theta^{2}}{Sin^{2}(\theta)}\right)\ with\ \beta>2$$

Quantiles: The inverse CDF (quantiles or chance functions, for α = 1 and β= [½, 1, 2, 4, 8 13)] is expressed as:

$$\frac{1}{F(p;\alpha,\beta)}=\alpha\left(\frac{p}{1-p}\right)^{1/\beta} \qquad [1.136]$$

1.34. Fisher distribution (F-distribution or Fisher–Snedecor)

Fisher distribution (Ronald Aylmer Fisher) is sometimes known as Fisher–Snedecor (George W. Snedecor) or F–Snedecor. It is a continuous distribution, which arises in null hypothesis distributions within tests of likelihood ratios or in the analysis of variance. Incidentally, we call it the F-test of variance. The F–Snedecor distribution is characterized by the random variable defined as the relationship between two independent variables, distributed according to Pearson distribution (χ^2 law):

$$\left(\frac{\tau_1}{\nu_1}\Big/\frac{\tau_2}{\nu_2}\right)=\left(\frac{\tau_1}{\nu_1}\right)\left(\frac{\nu_2}{\tau_2}\right);\nu_1\ and\ \nu_2\ \text{are the degrees of freedom of}\ \tau_1\ and\ \tau_2 \qquad [1.137]$$

In this case, we accept that a real RV with a density $f(\tau)$ which follows a Fisher–Snedecor probability distribution with (n_1, n_2) degrees of freedom (ν_1 and ν_2 integers > 0) if and only if its probability density, null for $\tau < 0$, is given, for $\tau > 0$, by the formula $f(\tau;\ \nu_1, \nu_2)$ below:

$$f(\tau)=\frac{\left(\dfrac{\nu_1.\tau}{\nu_1.\tau+\nu_2}\right)^{\frac{\nu_1}{2}}\times\left(1-\dfrac{\nu_1.\tau}{\nu_1.\tau+\nu_2}\right)^{\frac{\nu_1}{2}}}{\tau\times\beta\left(\dfrac{\nu_1}{2},\dfrac{\nu_2}{2}\right)} \qquad [1.138]$$

$$beta\left(\frac{\nu_1}{2},\frac{\nu_1}{2}\right)=\frac{\Gamma\left(\dfrac{\nu_1}{2}\right)\times\Gamma\left(\dfrac{\nu_2}{2}\right)}{\Gamma\left(\dfrac{\nu_1}{2}+\dfrac{\nu_2}{2}\right)}=3.142 \textit{ for } \nu_1=1 \textit{ and } \nu_2=1$$

where τ is a real number ≥ 0, (ν_1, ν_2) are its integers > 0, and *βeta* function is beta tabulated.

$$f\left(\tau;\nu_1,\nu_2\right)=\nu_1^{\frac{\nu_1}{2}}\times\nu_2^{\frac{\nu_2}{2}}\left(\frac{\Gamma\left(\dfrac{\nu_1+\nu_2}{2}\right)}{\Gamma\left(\dfrac{\nu_1}{2}\right)\cdot\Gamma\left(\dfrac{\nu_2}{2}\right)}\right)\times\frac{\tau^{\left(\frac{\nu_1}{2}-1\right)}}{\left(\tau\times\nu_1+\nu_2\right)^{\left(\frac{\nu_1+\nu_2}{2}\right)}} \qquad [1.139]$$

dF(τ, v1, v2) =

0.000
0.593
0.673
0.669
0.633
0.586
0.536
...

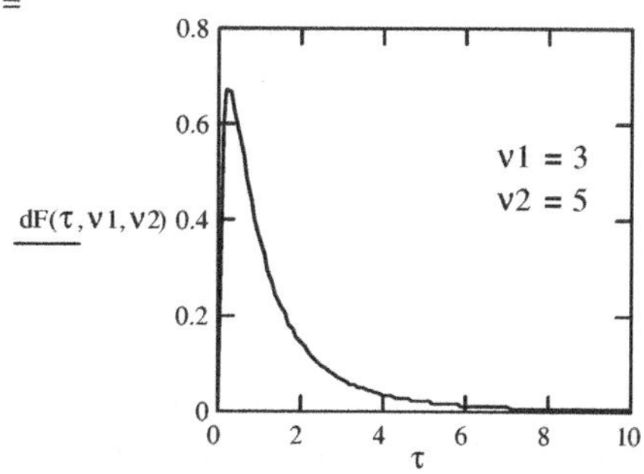

Figure 1.64. *Impression of F–Snedecor probability density function (1)*

The function of the *cumulative probability function* associated with F–Snedecor F(τ) is written as:

$$F(x)=\Upsilon\left(\frac{\nu_1.\tau}{\nu_1.\tau+\nu_2}\right)\left(\frac{\nu_1}{2},\frac{\nu_2}{2}\right) \qquad [1.140]$$

with $\Upsilon\left(\nu_1\cdot\tau\right)/\left(\nu_1\cdot\tau+\nu_2\right)$, a *regularized incomplete beta* function.

For $v_2 > 2$ the mean is written as:

$$\left(\frac{\nu_2}{\nu_2 - 2}\right)$$

and for $v_2 > 4$ it is:

$$\left(\frac{2.d_2^2\left(\nu_1 + \nu_2 - 2\right)}{\nu_1\left(\nu_2 - 2\right)^2 \times \left(\nu_2 - 4\right)}\right)$$

CDF of the F-distribution:

pF(τ, v1, v2) =

0
0.195
0.268
0.319
0.359
0.392
0.42
...

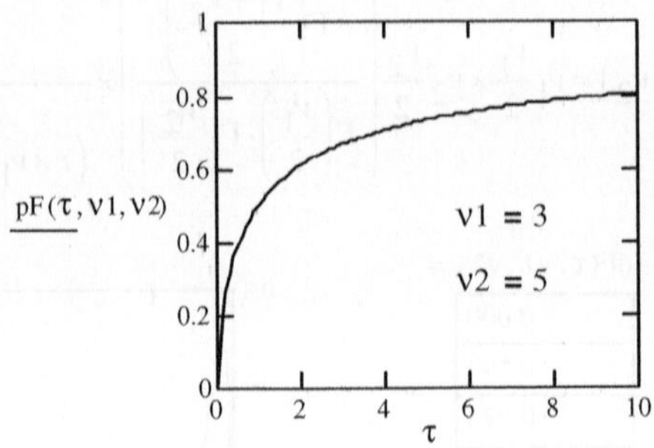

Figure 1.65. *Impression of the F–Snedecor cumulative distribution function*

Inverse CDF of F-distribution and vector of random variables distributed according to F-distribution are given in Figure 1.66:

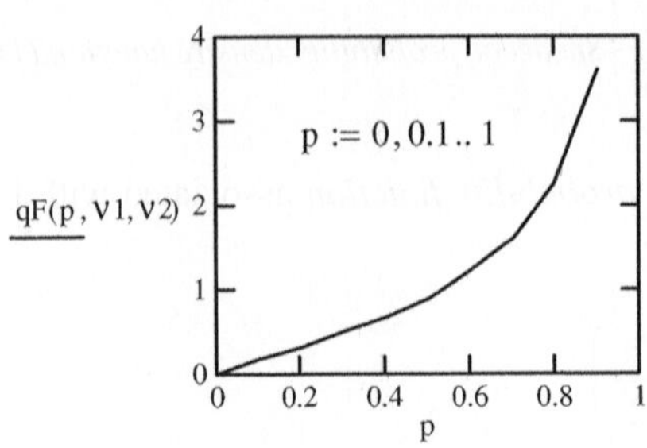

m = 7

$$rF(m, v1, v2) = \begin{pmatrix} 0.085 \\ 1.328 \\ 0.189 \\ 2.283 \\ 0.563 \\ 0.862 \\ 0.017 \end{pmatrix}$$

Figure 1.66. *Impression of the F–Snedecor inverse cumulative distribution function*

After having presented the main useful statistical distributions in mathematical modeling, we will conclude this chapter by presenting the tools necessary for the study of reliability of materials and structures.

1.35. Analysis of component lifespan (or survival)

In the mechanics of continuous surroundings, we bring the following distributions into play: Weibull, log-normal, gamma, normal, and above all, Birnbaum–Saunders. This last distribution remains the most representative in fatigue. The probability density of these types of law allows us to analyze the function of continuous (*non-monotone*) risk (failure). From Weibull distribution, we already know that with

$\beta > 1$, the risk function is said to be unimodal. Additionally, for $\beta \leq 1$, risk decreases in a monotone way (see Weibull distribution with two and three parameters).

The cumulative distribution of these laws gives rise to certain advantages. The formulation of this fact by an explicit cumulative probability function allows the expression of survival (lifespan or longevity) even when the data is truncated or censored: with the advantage going to experimental study. Censoring and truncating is often necessary in this type of work, due to the costs of experimentation (our metrology and quality control laboratory is not specifically dedicated to this type of work). The *lifespan* function is written as:

$$S(\tau) = 1 - F(\tau) = 1 \Big/ \left(1 + \left(\frac{\tau}{\alpha} \right)^{\beta} \right) \qquad [1.141]$$

The *risk function* is written as:

$$r(\tau) = \frac{f(\tau)}{S(\tau)} = \left(\left(\frac{\alpha}{\beta} \right) \times \left(\frac{\tau}{\alpha} \right)^{\beta - 1} \right) \Big/ \left(1 + \left(\frac{\tau}{\alpha} \right)^{\beta} \right) \qquad [1.142]$$

Log-logistic distribution comes into play quite often when modeling the precipitations and the flow of water in hydraulic domains and pluviometry. It is also in common usage among certain economists in finance (revenue from Canadian statistical data, for example where it takes the *premonitory* denomination of the *Fisk* distribution). We will not develop this distribution in this book due to lack of its application within fracture mechanics.

The *probability density function* of mass (distribution): We have simulated the trace of curves below by choosing variables, which do not correspond to experimental data. The objective here is purely pedagogical.

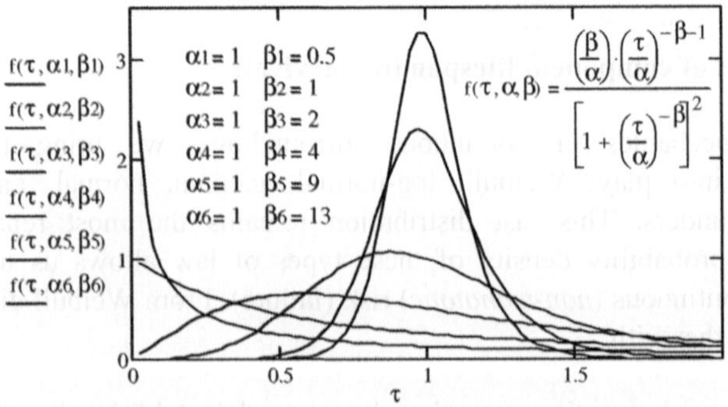

Figure 1.67. *Graphs showing the log-logistic distribution function for τ = [0 to2], α = 1, and β = [½, 1, 2, 4, 9 13)]*

Cumulative probability density function of mass:

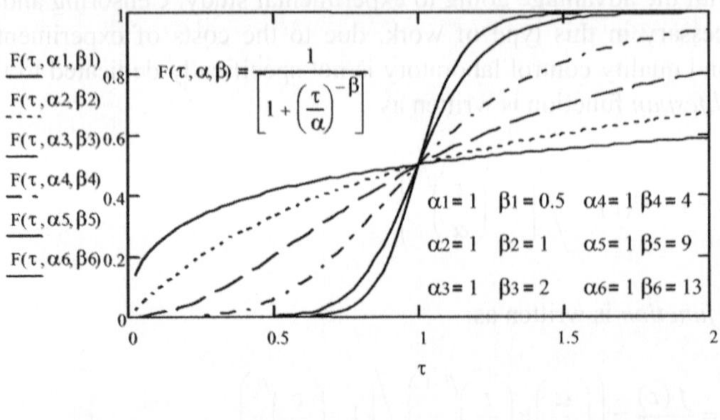

Figure 1.68. *Graph showing the cumulative distribution function of logistic distribution for τ = [0 to 2], α = 1, and β = [½, 1, 2, 4, 9 13]*

1.36. Partial conclusion of Chapter 1

We have seen that there are many statistical distributions and that sometimes these are related. The most important thing is not to simply choose a distribution and *burrow* into the application of everything that characterizes it, but rather to clearly note its applicability relative to case studies. In Chapter 3 of Volume 2 and

Chapter 5, Volume 2, we will deal with and comment on many different practical case studies which are essentially connected to the ingenious techniques of mechanics.

1.37. Bibliography

[BIR 68] BIRNBAUM Z.W., SAUNDERS S.C., "A probabilistic interpretation of miner's rule", *SIAM Journal of Applied Mathematics*, vol. 16, pp. 637–652, 1968.

[BIR 69a] BIRNBAUM Z.W., SAUNDERS S.C., "A new family of life distributions with applications to fatigue", *Journal of Applied Probability*, vol. 6, pp. 328–347, 1969.

[BIR 69b] BIRNBAUM Z.W., SAUNDERS S.C., "A new family of life distributions", *Journal of Applied Probability*, vol. 6, pp. 319–327, 1969.

[CHA 67] CHAKRAVARTI I.M., LAHA R.G., ROY J., *Handbook of Methods of Applied Statistics*, vol. I, John Wiley & Sons, pp. 392–394, 1967.

[CHA 72] CHAKRAVARTI L., ROY L., *Handbook of Mathematical Functions with Formulas, Graphs, and Mathematical Tables,* Dover Publications, New York, NY, 1972.

[COX 66] COX D.R., LEWIS PAW., *The Statistical Analysis of Series of Events*, John Wiley & Sons, Inc., New York, NY, 1966.

[COX 74] COX D.R., "Regression models and life tables", *Journal of the Royal Statistical Society*, vol. B 34, pp. 187–220, 1972.

[COX 84] COX D.R., OAKES D., *Analysis of Survival Data*, Chapman and Hall, London, New York, NY, 1984.

[CRO 74] CROW L.H., "Reliability analysis for complex repairable systems", in PROSCHAN F., SERFLING R.J. (eds.), *Reliability and Biometry*, SIAM, Philadelphia, pp. 379–410, 1974.

[EBI 97] EBELING C.E., *An Introduction to Reliability and Maintenability Engineering*, McGraw-Hill, 1997.

[ENG 82] ENGESVIK M.K., Analysis of uncertainties in fatigue capacity of welded joints, Report UR, Norwegian Institute of Technology, University of Trondheim, Norway, 1982.

[GRO 95] GROUS A., MUZEAU J.P., "Evaluation of the reliability of Cruciform structures connected by four welding processes with the aid of an integral damage indicator", *ICASP*, Civil Engineering Reliability and Risk Analysis. Blaise Pascal University, Clermont-Ferrand II, France, 1995.

[GRO 98] GROUS A., RECHO N., LASSEN T., LIEURADE H.P., "Caractéristiques mécaniques de fissuration et défaut initial dans les soudures d'angles en fonction du procédé de soudage" *Revue Mécanique Industrielle et Matériaux*, vol. 51, no. 1, Paris, France, April, 1998.

[GUM 54] GUMBEL E.J., *Statistical Theory of Extreme Values and Some Practical Applications*, National Bureau of Standards Applied Mathematics Series 33, U.S. Government Printing Office, Washington, D.C., 1954.

[GUM 08] GUMic Progiciel, Version 1.1., *Login_Entreprises*, Poitiers, France, 2008.

[KOV 97] KOVALENKO I.N., KUZNETSOV N.Y., PEGGY P.A., *Mathematical Theory of Reliability of Time Dependent Systems with Practical Applications*, John Wiley & Sons, Inc., New York, NY, 1997.

[MEE 75] MEEKER W.Q., NELSON W., "Optimum Accelerated Life-Tests for the Weibull and Extreme Value Distributions", *IEEE Transactions on Reliability*, vol. R-24, no. 5, pp. 321–322, 1975.

[MEE 85] MEEKER W.Q., HAHN G.J., "How to Plan an Accelerated Life Test – Some Practical Guidelines", *ASC Basic References In Quality Control: Statistical Techniques*, vol. 10, ASQC, Milwaukee, WI, 1985.

[MIL 78] MIL-STD-1635 (EC), Reliability Growth Testing, U.S. Government Printing Office, 1978.

[MIL 86] MIL-HDBK-217F, Reliability Prediction of Electronic Equipment, U.S. Government Printing Office, 1986.

[MOD 74] MOOD A., *Introduction to the Theory of Statistics*, 3rd ed., pp. 246–249, 1974.

[NEL 90] NELSON W., *Accelerated Testing,* John Wiley & Sons, Inc., New York, 1990.

[NIS 97] NIST 1997, *Engineering Statistics Handbook*, F-Distribution, 1997.

[O'CO 91] O'CONNOR P.D.T., *Practical Reliability Engineering,* 3rd ed., John Wiley & Sons, Inc., New York, NY, 1991.

[POR 98] PORES M., TOBIAS P.A., "How exact are 'Exact' exponential system MTBF confidence bounds?" *Proceedings of the Section on Physical and Engineering Sciences of the American Statistical Association*, 1998.

[TOB 85] TOBIAS P.A., TRINDALE D.C., *Applied Reliability,* 2nd ed., Chapman and Hall, London, New York, NY, 1995.

[TOV 01] TOVO R., "On the fatigue reliability evaluation of structural components under service loading", *International Journal of Fatigue*, vol. 23, pp. 587–598, 2001.

[WEI 51] WEIBULL W., "A statistical distribution function of wide applicability", *Journal Applied Mechanics-Transactions ASME,* vol. 18, no. 3, pp. 293–297, 1951.

Chapter 2

Estimates, Testing Adjustments, and Testing the Adequacy of Statistical Distributions

2.1. Introduction to assessment and statistical tests

Often, we choose a distribution based on the fact that it is the most well-known distribution adapted to the physical phenomena to be considered. However, there is nothing that predisposes a certain distribution toward being systematically designed to deal with a certain phenomenon. Therefore, we must turn to adequacy tests (concordance) to accept a certain hypothesis, or declare it null.

In fact, these tests are there to attempt to respond to the probability, which is supposed to be represented by the distribution and set out by the distribution chosen.

Is there still a correct approach to follow? Yes and no.

Yes: The probability laws are very "solid" mathematically and well anchored in experimental (at times, empirical) life. They are less expensive in terms of calculation time (contrary to certain simulations such as the Monte Carlo simulation). These tests are used as concordance tests to measure the standard deviation between the empirical distribution function and the theoretical distribution function.

No: It is sometimes dangerous to "force" the applicability of a distribution and to "make do with" an "inadequate" test to prove something. This is dangerous in terms of security and shows low levels of mathematical rigor.

We will present the required tools for studying probability in reliability and in quality control. Their use targets the estimates of parameters of probability distributions, primarily chosen based on conformity tests. Statistical estimates allow us to characterize the distribution of a dataset or of a part of this population, shown as $\{\tau_1,\ldots, \tau_n\}$. Worked out based on a series of realizations of a random variable, they constitute random variables that we can use to specify probability laws. The probabilistic modeling of physical phenomena, while rigorously taking on experimental data, is an inescapable reality. It requires the cooperation of multi-disciplinary teams, as the terms in use can sometimes appear unclear due to the concepts behind them seeming obscure. Additionally, the programs, as reliable as they are robust, cannot come up with ideas themselves. They only calculate what we want them to send back. It is irritating that manipulating rigorous probabilistic concepts, without understanding the mechanisms that they put into practice, is a concept, which is taking over.

In statistical inference, we usually associate statistical tests with estimates. The objective of these two fields is to generalize the information from the sample to the dataset. The purpose of this is to make the characteristics of the dataset known to then establish the most representative model possible and to make the best decisions pertaining to this. The statistical test is meant to judge the dependability of a hypothesis, which has already been formulated based on a distribution hypothesis. This is very important because the decision will affect other things and this is why the hypotheses must be tested with the utmost mathematical rigor. The estimates supply a certain evaluation of the parameter of a dataset. The estimate can give information about the mean, the variance, etc. In sum, it gives information about the main characteristics of the chosen distribution and is used according to information from the hypothesis. The test and the estimate are equally based on statistical evidence from a sample.

We will present a few salient elements of the estimate and tests applied in material and structure reliability, as well as in quality control. Then, we will propose a few case studies, exclusively for pedagogic reasons.

2.1.1. *Estimation of parameters of a distribution*

Overview: Estimation is generally subdivided into two distinct domains: interval estimation and punctual estimation. Punctual estimation evaluates a parameter of the dataset, which stems from statistics measured on a sample. Interval estimation allows us to determine a certain confidence interval, whose probability would contain the **true value** (see VIM = Vocabulaire International de Métrologie–International Vocabulary for Metrology) with this parameter [GRO 09, GRO 11].

For us to calculate an indication of the *true* value of a parameter of the dataset, with the help of punctual estimation, we must first determine the requirements which have led us to believe that we have made the correct choice; a choice which constitutes a reliable and precise indication of the probabilistic representation of a model. The problem is that the estimators themselves constitute random variables, and that some of them are also interdependent of the mean or of the variance of the estimator in question. Before dealing with this, we will present the main properties of the estimators.

i) The *arithmetic mean* is the convergent estimator of the mean of variable (τ). This appears as:

$$\mu_\tau = \frac{1}{n} \times \sum_{i=1}^{n} \tau_i \qquad [2.1]$$

where μ_τ follows a distribution of mean $E(\tau)$ and of variance $V(\tau)/n$.

ii) The *quadratic mean* is the convergent estimator of the variance of (τ) whose expression is:

$$VA(\tau) = \sigma_\tau^2 = \frac{1}{n} \times \sum_{i=1}^{n} (\tau_i - \mu_\tau)^2 \;\; or \;\; \sigma_\tau = \sqrt{\frac{1}{n} \times \sum_{i=1}^{n} (\tau_i - \mu_\tau)^2} \qquad [2.2]$$

The mean and the variance of the distribution are:

$$\left(\frac{n-1}{n}\right) VA(\tau) \;\; and \;\; \left(\frac{n-1}{n}\right)^2 \times \frac{E\left[\tau - E(\tau)\right]^4 - VA(\tau)^2}{n} + 2\left(\frac{n-1}{n^3}\right) \times VA(\tau)^2$$

A **bias-free** estimator of the variance of X is $n.\sigma_\tau^2/(n-1)$. We must note that μ_τ and σ_τ^2 are two independent variables.

iii) The **covariance** between two series of observations $\{x_1, \ldots, x_n\}$ and $\{y_1, \ldots, y_n\}$ can be estimated by:

$$\sigma_{xy}^2 = \frac{1}{n} \times \sum_{i=1}^{n} (x_i - \mu_x)(y_i - \mu_x) \qquad [2.3]$$

2.1.2. *Estimation by confidence interval*

The confidence interval of risk (α) **specifies the interval in which a statistical estimator is situated with a probability equal to** $(1-\alpha)$. The characteristics of a confidence interval depend on the probability law postulated in the data and the known parameters postulated with certainty of this law.

2.1.2.1. *Confidence interval of the mean of a Gaussian distribution of known variance*

If $U_{\alpha/2}$ indicates the value of a normal, centered, reduced variable U such as $P\left\{ |U| \geq U_{\alpha/2} \right\} = \left(\frac{\alpha}{2}\right)$, the confidence interval of risk α, is:

$$\left\{\left(\mu_\tau - U_{\alpha/2}\right)\left(\frac{\overline{VA(\tau)}}{n}\right)\right\} \leq E(\tau) \leq \left\{\left(\mu_\tau + U_{\alpha/2}\right)\left(\frac{\overline{VA(\tau)}}{n}\right)\right\} \quad [2.4]$$

2.1.2.2. *Confidence interval of the Gaussian mean of unknown variance*

If $t_{\alpha/2}$ indicates the value of a T variable following a Student distribution to $(n-1)$ degrees of freedom, such as $P\left\{|U| \geq U_{\alpha/2}\right\} = \left(\frac{\alpha}{2}\right)$, the confidence interval α, of risk, is:

$$\left\{\left(\mu_\tau - t_{\alpha/2}\right)\left(\frac{\overline{\sigma_\tau}}{n}\right)\right\} \leq E(\tau) \leq \left\{\left(\mu_\tau + t_{\alpha/2}\right)\left(\frac{\overline{\sigma_\tau}}{n}\right)\right\} \quad [2.5]$$

2.1.2.3. *Confidence interval of the variance of a normal distribution*

If χ_1^2 and χ_2^2 indicate the values of a χ^2 variable following chi square distribution to $(\nu = n-1)$ degrees of freedom, respectively, such as:

$$P\left\{\chi^2 \leq \chi_1^2\right\} = \frac{\alpha}{2} \quad and \quad P\left\{\chi^2 \geq \chi_2^2\right\} = \frac{\alpha}{2}$$

The confidence interval of risk (α) would be written:

$$\left\{\frac{n \times \sigma_\tau}{\chi_2^2} \leq \mathrm{VA}(\tau) \leq \frac{n \times \sigma_\tau}{\chi_1^2}\right\} \quad [2.6]$$

2.1.3. *Properties of an estimator with and without bias*

We are dealing with a random variable (RV) (X) of distribution characterized by a parameter θ to be estimated. The bias-free mathematical condition stipulates that for a single statistic, $\hat{\theta}$ must be considered an estimator without bias of parameter (θ), the mean of distribution of $\hat{\theta}$ has to be equal to the value of $\hat{\theta}$ and the following must be written:

$$E\left(\hat{\theta}\right)=\theta \text{, with } \left(\hat{\theta}\right) \textit{ Estimator without bias}$$

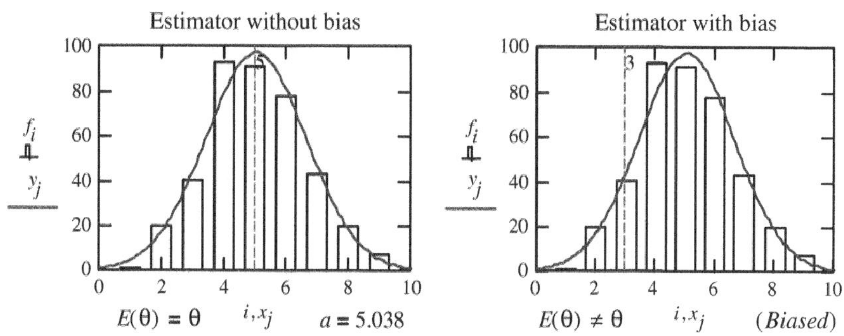

Figure 2.1. *Representation of the properties of the biased estimator and the estimator without bias*

Demonstration using the example of the mean μ and the variance s^2 where the standard deviation is (σ)

We demonstrate: $E(\mu)=\mu$, *with* $\mu=\dfrac{\sum_{i=1}^{n} x_i}{N}$

Solution:

$$\begin{cases} \textit{Knowing that } E(\mu)=E\dfrac{\left(\sum_{i=1}^{n} x_i\right)}{N}=\dfrac{E\left(\sum_{i=1}^{n} x_i\right)}{N}=\dfrac{\sum_{i=1}^{n} E(x_i)}{N} \\ \textit{As } E(x_i)\rightarrow E(\mu)=\dfrac{\sum_{i=1}^{n} E(x_i)}{N}=\dfrac{\sum_{i=1}^{n} \mu}{N}=\dfrac{N}{N}\times\mu=\mu \end{cases} (QED), \textit{what it was necessary to demonstrate}$$

We thus conclude that $\overline{x}$ is an estimator without bias of (μ).

Question: *Are we demonstrating that* $E\left(s^2\right)=\sigma^2$ or that $E(s)=\sigma$

Solution: Knowing that: $s^2=\sum_{i=1}^{k}(x_i-\overline{x})\Big/(N-1)$ we propose: $\sum_{i=1}^{k}(x_i-\overline{x})$

As $\sum_{i=1}^{k}\left[(x_i-\mu)-(\overline{x}-\mu)^2\right]$ we propose:

$$E\left(s^2\right)=E\left(\sum_{i=1}^{n}\left[(x_i-\mu)-(\overline{x}-\mu)^2\right]\Big/(n-1)\right)\Rightarrow$$

$$E\left(s^2\right)=\left(\frac{1}{n-1}\right)\cdot E\left\{\sum_{i=1}^{n}\left[(x_i-\mu)^2-2(x_i-\mu)(\overline{x}-\mu)+(\overline{x}-\mu)^2\right]\right\}$$

$$E\left(s^2\right)=\left(\frac{1}{n-1}\right)\cdot E\left\{\sum_{i=1}^{k}\left[(x_i-\mu)^2-2(\overline{x}-\mu)\sum_{i=1}^{n}(x_i-\mu)+\sum_{i=1}^{n}(\overline{x}-\mu)^2\right]\right\}$$

$$E\left(s^2\right)=\left(\frac{1}{n-1}\right)\cdot E\left\{\sum_{i=1}^{n}\left[(x_i-\mu)^2-2(\overline{x}-\mu)\times n\times(\overline{x}-\mu)+n\times(\overline{x}-\mu)^2\right]\right\}$$

$$E\left(s^2\right)=\left(\frac{1}{n-1}\right)\cdot E\left\{\sum_{i=1}^{n}\left[(x_i-\mu)^2-2n(\overline{x}-\mu)+n\times(\overline{x}-\mu)^2\right]\right\}$$

$$E\left(s^2\right)=\left(\frac{1}{n-1}\right)\cdot E\left\{\sum_{i=1}^{n}\left[(x_i-\mu)^2-n\times(\overline{x}-\mu)^2\right]\right\}$$

$$E\left(s^2\right)=\left(\frac{1}{n-1}\right)\cdot E\left\{\sum_{i=1}^{n}\left[(x_i-\mu)^2-\frac{n}{n-1}\times E(\overline{x}-\mu)^2\right]\right\}$$

Already knowing that $E(x_i-\mu)^2=\sigma^2$ *as well as* $E(\overline{x}-\mu)^2=\sigma^2/n$

$$We\ deduce\ that\ E\left(s^2\right)=\left(\frac{1}{n-1}\right)\times\sum_{i=1}^{n}\sigma^2-\left(\frac{1}{n-1}\right)\times\left(\frac{\sigma^2}{n}\right)=$$

$$=\left(\frac{n\times\sigma^2}{n-1}\right)-\left(\frac{\sigma^2}{n-1}\right)=\sigma^2\left(\frac{n-1}{n-1}\right)=\sigma^2\ \to QED$$

We conclude, by demonstrating, that s^2 is an estimator that is bias-free of σ^2. By analogy, s is, in reality, an estimator of standard deviation (σ).

Application: Gaussian distribution for i = [1; 1; 20; 40; 93; 91; 78; 43; 20; 7; 1]

$$N=\sum_i f_i=395;\ \overline{x}=\frac{\sum_i i\times f_i}{N}=5.038\ and\ \sigma=\sqrt{\frac{\sum_i f_i\left(i-\overline{x}\right)^2}{N-1}}=1.625$$

$$For\ i=0\ to\ 100;\ x_j=\left(\frac{j}{10}\right)and\ Y_j=\frac{1}{\sigma\sqrt{2\pi}}Exp\left(-\frac{\left(x_j-\overline{x}\right)^2}{2\sigma^2}\right)\times N\ for\ cnorm(-1)=0.159$$

$$For\ \overline{x}=\mu=0\ and\ \sigma=1\to\int_{-1000}^{-1}\frac{1}{\sigma\sqrt{2\pi}}\times Exp\left(-\frac{\left(x_j-\overline{x}\right)^2}{2\sigma^2}\right)dx=0.158655$$

Because estimators are fundamentally based on observations from a sample, it is logical that they may also be random variables inherent to a distribution with an appropriate mean, variance, and standard deviation. Moreover, it is often by using a first-order mean that one is capable of detecting bias. Consequently, the importance of the mean of the estimator allows a clear reading of the efficacy of this estimator. In the study of fracture by fatigue, we often have to work with a weak variance, or even a variance containing bias. However, the better estimators are still the linear, centered estimators.

In addition, the estimators possess asymptotic properties, which must be respected. Many methods, including digital and graphical methods, that can be used to estimate the parameters of a given probability distribution. We present the three most important methods named below:

i) Method of Moments (MM)

ii) Method of Maximum Likelihood (MLE)

iii) Moving Least-Squares (MLS) Method

Attention: In the study of *convergence*, in section 3.5 volume 2, we will present a case study relative to the Weibull distribution using the Miner function for the adjustment of the MLS of nonlinear functions. The Miner method refers to the last values calculated when the maximum number of iterations toward a solution would have been attained without fulfilling the convergence.

2.2. Method of moments

The MM is equal to the moments of the sample of parameter estimation. These methods have the advantage of being simple, but their unavailability presents a non-negligible disadvantage. Their optimization remains occasionally subject to caution. For this reason, we prefer the *MLE* and *MLS method* for the estimator.

2.3. Method of maximum likelihood

The estimation of maximum likelihood begins with the mathematical expression known as a function of likelihood of the data within a sample. The probability of a collection of data is the probability of obtaining this particular dataset given the model of probability chosen. This expression contains unknown parameters. The values of the parameter, which *maximize the likelihood* of the sample are known as the estimations of the maximum likelihood.

The advantages of this method are:

– Maximum likelihood gives a coherent approach to the estimation problems of the parameters for a wide variety of estimation situations. Let us cite, for example, the reliability analysis of the data censored in different models of censoring.

– Good mathematical properties and their optimality because:

- They become estimators of impartial minimal variance to the measure of increasing the size of the sample. The mean value of the estimation of parameters will theoretically be equal to the value of the dataset. By minimal variance, we mean that the estimator has a small difference, which is where the narrow confidence interval of all the estimators comes from.

- They have an approximately normal distribution, whose variances from the sample are used to generate confidence intervals and hypothesis tests for the parameters.

In the literature, numerous programs give excellent algorithms of the estimations of the maximum likelihood for the most commonly used distributions. This helps to attenuate the complexity of the calculation of estimation.

The disadvantages of this method are that:

– The probability equations must be specifically elaborated on for a given distribution of the problem inherent to estimation. Mathematics is non-trivial in terms of confidence intervals and desirable parameters.

– The numerical estimation is normally non-trivial. The estimations of maximum likelihood can be highly biased when the sample size is small. This shows that the properties of optimality can cause problems for small sample sizes, thus showing the importance of the number of experimental data or simulations included. The final thing to note is that methods of maximum likelihood are sometimes sensitive to the choice of the initial values.

2.3.1. *Estimation of maximum likelihood*

The MLE is a powerful and precise method for large samples. To apply it, one must initially write the mathematical expression, which correctly represents the model. This expression contains the parameters of the unknown model. The values of the parameters, which maximize the probability of the sample, are known as the estimations of maximum likelihood. In fact, the ***EML = Estimation of Maximum Likelihood*** is a procedure of analytical maximization. It applies to all *censored* or *multi-censored* forms of data. It is possible to use this technique in fatigue and to estimate the parameters of a model of acceleration at the same time, as one can with parameters of lifespan distribution. Furthermore, the functions of likelihood generally have known properties of large sample size, which are:

– estimators that are non-biased in terms of minimum variance according to the sample size;

– close to normal distribution used to generate confidence limits;

– functions of likelihood used to test hypotheses on the models;

– With small samples, the ML is perhaps not very *persuasive*, but it can generate a line, which can be found above or below the given points. There are considerable disadvantages to ML which are:

i) With a small number of failures (<5), MLE can be highly biased and properties of large samples no longer apply.

ii) The calculation of MLE requires informatics tools (programs) which are specialized for complex problems of nonlinear equations.

2.3.2. *Probability equation of reliability-censored data*

$f(t)$ is a distribution function and $F(t)$, is its cumulative probability function for the lifespan distribution model chosen. We must clearly note that these are functions of (τ) and are parameters unknown to the model. The probability function for **censored tests** is shown below:

$$\ell = \kappa \times \left(\prod_{i=1}^{r} f(\tau_i) \right) \times 1 - F(T)^{n-r} \; ; \; \kappa = cst \tag{2.7}$$

This is how to organize and read censored tests:

N components to test $\qquad$ *$n - \Sigma r_i \; (T_k = T)$*

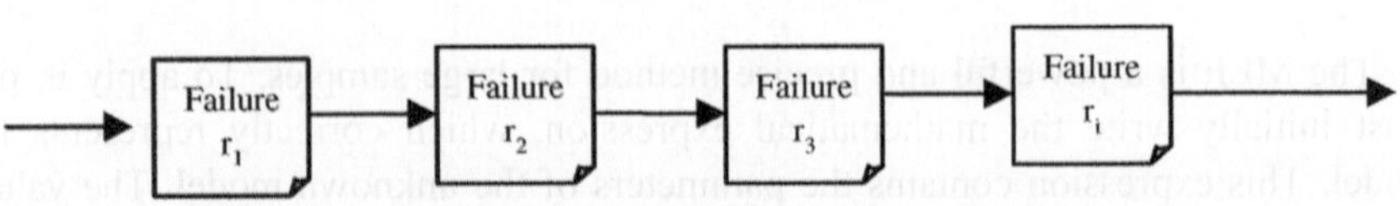

Figure 2.2. *Organization of components in series (censored test)*

In the case of **non-censored tests,** probability reduces itself only to the product of densities, each one evaluated at a moment of fracture. For type 2 censored tests, the probability function is:

$$\ell = \kappa \times \left(\prod_{i=1}^{k} F\left(T_i - F\left(T_{i-1}\right)\right)^{r_i} \right) \times \left(1 - F(T)\right)^{n - \sum_{i=1}^{k} r_i} \; ; \; F(T_0) \textit{ defined to } 0 \tag{2.8}$$

T resolve this MLE problem, we define the partial derivatives of ln (l). The equation for the **exponential model,** for example, follows:

$$\ell = \kappa \times \lambda^r \times \text{Exp}\left\{ -\left(\lambda \sum_{i=1}^{r} \tau_i \right) \right\} \times \text{Exp}\left\{ -\lambda (n-r) \right\} T \textit{ hence} \rightarrow$$

$$\rightarrow \ln(\ell) = \ln(\kappa) + r \ln(\lambda) - \lambda \sum_{i=1}^{r} \tau_i - \lambda (n-r) T \tag{2.9}$$

The failure rate's MLE (or the rate of repair) in the expression of power, (i.e. exponential) gives the total number of failures observed, divided by the total time of censored testing.

$$\frac{\partial \ln(\ell)}{\partial \lambda} = \frac{r}{\lambda} - \sum_{i=1}^{r} \tau_i - \lambda(n-r)T = 0 \ so \ \hat{\lambda} = \frac{r}{\sum_{i=1}^{r} \tau_i - \lambda(n-r)T} \quad [2.10]$$

2.3.3. *Punctual estimation of exponential law*

The proposed estimation (below) focuses on exponential distribution where we use (τ_f) to indicate total, functional lifespan and (κ) to indicate the number of failures observed; note the following, taken from the literature:

$$\left\{ \hat{\lambda} = \left(\frac{\kappa}{\tau_f} \right) \ or \ \hat{\omega} = \left(\frac{\tau_f}{\kappa} \right) \right\} \quad [2.11]$$

We must note that for (κ) = 0, the literature suggests using the following figure for estimation $\hat{\omega}$, a value which corresponds to a probability of ½ to get:

$$\left\{ \hat{\lambda} = \left(\frac{0.69}{\tau_f} \right) \ or \ \hat{\omega}_0 = \left(\frac{\tau_f}{0.69} \right) \right\} \quad [2.12]$$

We observe two distinct cases of tests:

– if the test is carried out by replacing failing components, on the collection of (n) parts, during the time interval (τ), we will have = $nx(\tau)$;

– if the test is carried out without replacing failing components on the collection of (n) parts, during the time interval (τ), we will have two distinct cases to observe:

CASE 1:– Test deliberately stopped (*censored*) after a certain number of failures F, previously fixed:

$$\tau_f = (n+1-F) \times \tau_F + \sum_{i=1}^{F-1} (\tau_i) \quad [2.13]$$

CASE 2:– Deliberately stopped tests (*truncated*) after a certain number of fixed tests (τ), where we noticed (κ) failures.

$$\tau_f^{truncated} = (n-\kappa) \times \tau + \sum_{i=1}^{\kappa} (\tau_i) \quad [2.14]$$

Confidence Interval: For *censored tests*, at the confidence threshold $\{1-(\alpha_1+\alpha_1)\}$:

$$\left\{\left(\frac{2\times\tau_f}{\chi^2_{(2R;\,\alpha_2)}}\right)\prec\omega\prec\left(\frac{2\times\tau_f}{\chi^2_{(2R;\,1\text{-}\alpha_1)}}\right)\right\} \qquad [2.15]$$

$\chi^2_{(2R;\,\alpha_2)}$, represents the χ^2 variable to $2R$ degrees of freedom, which must exceed probability (α). We will read the tabulated values in the appendix. For **censored tests**, at the confidence threshold $\{1-(\alpha_1+\alpha_1)\}$ we postulate:

$$\left\{\frac{2\times\tau_f}{\chi^2_{[(2\kappa+1);\,\alpha_2]}}\prec\omega\prec\frac{2\times\tau_f}{\chi^2_{[(2\kappa;\,1\text{-}\alpha_1)]}}\right\} \qquad [2.16]$$

2.3.4. *Estimation of the Weibull distribution*

The estimation proposed here deals with the Weibull distribution with two parameters (β and η). We postulate $\lambda=1/(\eta\times\beta)$ to then estimate the parameters, which constitute the following solution to the system below:

$$\left\{\begin{array}{l}\left(\dfrac{n}{\beta}\right)+\displaystyle\sum_{i=1}^{n}\ln\tau_i-\lambda\times\sum_{i=1}^{n}\left(\tau_i^{\beta}\times\ln\tau_i\right)=0\\[2ex]\left(\dfrac{n}{\lambda}\right)-\displaystyle\sum_{i=1}^{n}\left(\tau_i^{\beta}\right)=0\end{array}\right\} \qquad [2.17]$$

Usually, the solution to this system is carried out by successive approximations, solving the calculation of which appears below:

$$\beta=f(\beta)=\frac{n\times\sum_{i=1}^{n}\left(\tau_i^{\beta}\right)}{n\times\sum_{i=1}^{n}\left(\tau_i^{\beta}\right)\times\ln(\tau_i)\times\sum_{i=1}^{n}\ln(\tau_i)\times\sum_{i=1}^{n}\left(\tau_i^{\beta}\right)} \qquad [2.18]$$

$$\beta_{(i+1)}=\frac{\beta_{(i+1)}+f(\beta_i)}{2}\quad \textit{and}\quad \eta=\frac{\sum_{i=1}^{n}\left(\tau_i^{\beta}\right)^{\left(\frac{1}{\beta}\right)}}{n} \qquad [2.19]$$

For the convenience of calculation (which can often be fastidious and sometimes complex), it is preferable to use the graphical estimation using a functional scale (See Appendix A.1 Volume 2). The confidence interval for (η) is easily determined if (β) is known.

The expression, which comes from the literature is shown below:

$$\left\{ \left(\frac{2 \times \sum_{i=1}^{n} \left(\tau_i^{\beta} \right)}{\chi^2_{\left((2n+2);\ \frac{\alpha}{2} \right)}} \right)^{\left(\frac{1}{\beta} \right)} \prec \eta \prec \left(\frac{2 \times \sum_{i=1}^{n} \left(\tau_i^{\beta} \right)}{\chi^2_{\left((2n);\ \left(1-\frac{\alpha}{2} \right) \right)}} \right)^{\left(\frac{1}{\beta} \right)} \right\} \qquad [2.20]$$

We will read this as: $\chi^2_{(\nu;\ \alpha)}$, which is the variable of χ^2 to (ν) degrees of freedom with a probability (α) to be exceeded.

2.3.5. *Punctual estimation of normal distribution*

The punctual estimator of the mean is: $\hat{\mu}_{mean} = m = \left(\frac{1}{n} \right) \times \sum_{i=1}^{n} \tau_i$

The punctual estimator of the variance is:

$$\hat{V} = \hat{\sigma}^2 = \hat{s}^2 = \left(\frac{1}{n-1} \right) \times \sum_{i=1}^{n} (\tau_i - m)^2 = \left(\frac{1}{n-1} \right) \times \sum_{i=1}^{n} \left(\tau_i^2 - n \times m^2 \right) \qquad [2.21]$$

We should note that the estimator of the standard deviation is written equally as (s). The standard deviation varies rigorously from [1.253s] for $n = 2$ to s for the infinite.

Symmetrical bilateral confidence interval at the confidence level ($1 - \alpha$)

The symmetrical, bilateral confidence interval for the mean is:

$$\left\{ m - \left(\tau_{n-1;\ \alpha} \right) \cdot \frac{s}{\sqrt{n}} \right\} \prec \mu \prec \left\{ m + \left(\tau_{n-1;\ \alpha} \right) \cdot \frac{s}{\sqrt{n}} \right\} \qquad [2.22]$$

where $\tau_{(n-1;\,\alpha)}$ is the value of the Student variable to $(n-1)$ degrees of freedom, for a symmetrical risk (α). The bilateral symmetrical confidence interval for variance is written:

$$\frac{(n-1)\times s^2}{\chi^2_{\left((n-1);\,\frac{\alpha}{2}\right)}} \prec \sigma^2 \prec \frac{(n-1)\times s^2}{\chi^2_{\left((n-1);\,1-\frac{\alpha}{2}\right)}} \qquad [2.23]$$

$\chi^2_{(n-1;\,\alpha/2)}$, is the value of the χ^2 variable to $(n-1)$ degrees of freedom

Estimation of log–normal distribution (or Galton's): In mechanical reliability, the domain is never negative. Consequently, there is no limitation concerning the employment of the Galton distribution. We have often used it in fracture (by fatigue) mechanics. It allows a clear representation of phenomena of fatigue of structures assembled by soldering [GRO 94, GRO 98]. As we can see from reading the graph of this distribution (see Chapter 1), the failure rate increases at the beginning of life. The distribution is dissymmetrical.

Estimation of Gamma distribution (Erlang): We will proceed to the estimation either by the MM, or by the MLE. We will first present the formulae and a simple example of numerical application. Then, we will suggest a case study in Chapter 8, Volume 2.

a) Method of moments

$$\left\{\hat{\lambda} = \left(\frac{m}{s^2}\right) \text{ and } \hat{\kappa} = \left(\frac{m^2}{s^2}\right) \text{ with } m = \left(\frac{1}{n}\right)\sum_{i=1}^{n}\tau_i \text{ and } s^2 = \left(\frac{1}{n-1}\right)\sum_{i=1}^{n}\left(\tau_i - m\right)^2\right\} \qquad [2.24]$$

b) Method of maximum likelihood

Calculation of $\hat{\kappa}$: It is best to resolve the equation in the following way before doing anything else:

$$\left\{\ln\kappa - f(\kappa-1)\right\} = \ln\frac{\sum_{i=1}^{n}\tau_i}{n} - \frac{\sum_{i=1}^{n}\ln\tau_i}{n} \qquad [2.25]$$

With $f(\kappa-1) = \Gamma'(\kappa)/\Gamma(\kappa)$ which is, in fact, the logarithmic derivative the relationship between the arithmetic mean m and the geometric mean $\rho = \sqrt[n]{\prod_{i=1}^{n}\tau_i}$. We can see that the term to the left of the equation [2.25] is a decreasing function of κ, hence we have:

$$\text{For: } \ln\left(\frac{m}{\rho}\right) \succ 0.379\,6\,;\ \left(\frac{m}{\rho}\right) \succ 1.461\,6 \ \textit{So}\ Ln(\ell) = -0.549539 - 1.138058\ln(\kappa);$$

$$\text{Hence}\begin{cases}\hat{\kappa} = 0.617007 \times \ln\left(\frac{m}{\rho}\right)^{-0.8786896} \quad ; \ \ln\left(\frac{m}{\rho}\right) \prec 0.3796 \ \rightarrow \textit{So}\ Ln(\ell) = \\ = -\ 0.580317 - 1.049810 \times \ln(\kappa)\ \textit{hence}\ \hat{\kappa} = 0.575346 \times \ln\left(\frac{m}{\rho}\right)^{-0.952653}\end{cases}$$

Calculation of $\hat{\lambda}$: $\hat{\lambda} = (n \times \hat{\kappa}) \Big/ \sum_{i=1}^{n} \tau_i = \left(\frac{\hat{\kappa}}{m}\right)$ [2.26]

2.4. Moving least-squares method

The MLS method, as its name indicates, aims to minimize quadratic error between an estimator and the estimated parameter.

The advantages of this method are:

– It can be applied more generally than the maximum of likelihood. To put it another way, if the program used allows a nonlinear model case and if there is the possibility of specifying the function of probability studied. One can generate the *least squares* from estimations for this distribution. That will allow you to obtain reasonable estimations for distributions, even if the program does not supply estimations of maximum likelihood.

The disadvantages of this method are:

– It is not easily applicable to censored data.

– It is generally considered to have optimality properties, which are less desirable than those of the maximum likelihood method.

– It can be quite sensitive to the choice of initial values (hypothesis).

Contrary to linear regression, there are very few limitations in parameters that can be used in the model of nonlinear regression. The way in which unknown parameters are estimated is identical to that of linear regression of least squares. As its name indicates, a nonlinear model is different from a simple model whose expression of the regression function is:

$$y = f\left(\vec{x};\vec{\beta}\right) + \varepsilon \quad \leftarrow \left(\textit{regression function}\right) \qquad [2.27]$$

– In which the functional part of the model is nonlinear compared to unknown parameters β_0, β_1, β_2, β_n, etc. …

– The MLS method is used to estimate the values of unknown parameters.

Due to the manner in which the unknown parameters of the function are generally estimated, it is often easier to work with models, which respond to two supplementary criteria:

– The function is *smooth* compared to unknown parameters.

– The *least-squares* criterion **(LSC)** is used to obtain estimations of parameters with a unique solution.

2.4.1. *General criterion: the LSC*

In the case of estimation by the least-squares method (*LS* = *Straight Line*), the values of unknown parameters, β_1, β_2, …, β_n in the regression function $f\left(\vec{x};\vec{\beta}\right)$ are estimated by numerical values to find the parameters which *minimize the sum of the squares of separation between the observed responses and the functional part of the model.* Mathematically, the minimum criterion (sum of squares) is:

$$\Psi = \sum_{i=1}^{n}\left(y_i - f\left(\vec{x};\vec{\beta}\right)\right)^2 \qquad [2.28]$$

The factors β_1, β_2, … β_n are dealt with by using optimization variables and the values of predictive variables, x_1, x_2, … x_n are the coefficients. To underline the fact that the estimations of parameter values are not the same as the same values of the parameters, the estimations are indicated by $\hat{\beta}_0, \hat{\beta}_1, \hat{\beta}_2, \ldots, \hat{\beta}_n$. For linear models, the minimization of squares is habitually done using analytical calculation. For nonlinear models, minimization has to be, and is, generally carried out using iterative numerical algorithms. To illustrate the line of regression, we will consider the following linear model:

$$y = \beta_0 + x\beta_1 + \varepsilon \qquad [2.29]$$

For this model, the estimations by the least squares of parameters are calculated by minimizing:

$$\Psi = \sum_{i=1}^{n}\left[y_i - \left(\hat{\beta}_0 + x\hat{\beta}_1\right)\right]^2 \qquad [2.30]$$

To this effect, it is appropriate to:

– consider the partial derivatives of Ψ compared to $\hat{\beta}_0$ and $\hat{\beta}_1$;

– regulate each partial derivative to zero;

– resolve the system to two equations to two unknowns.

In terms of *efficiency*, estimators for parameters look like:

$$\hat{\beta}_0 = \overline{y} - \overline{x}\hat{\beta}_1 \ \ and \ \ \hat{\beta}_1 = \sum_{i=1}^{n}\left(x_i - \overline{x}\right)\left(y_i - \overline{y}\right) \Big/ \sum_{i=1}^{n}\left(x_i - \overline{x}\right)^2 \qquad [2.31]$$

These formulae show that estimators are not independent from each other, except for $\overline{x} = 0$. The formula of the estimator of interception depends directly on the value of the estimator of the gradient, except in the case where the second term in the formula of $\hat{\beta}_0$ decreases due to being multiplied by zero. That indicates that if the estimation of the gradient separates itself considerably from the *true gradient*, the estimation of interception will have a tendency to break away from its real value even more as well. This lack of independence of estimators of parameters or more precisely, the correlation of estimators of parameters becomes important during the calculation of the uncertainty of values predicted by the model.

Although the above formulae do not only apply to the linear model, the relationship between the estimators of the parameter is analogous with the most complex models (*nonlinear models*). The comparison between the LS line and the true line using the example of the practical application of a linear regression is presented in this way as an example:

Initial data: Statistical function of the linear regression of X–Y data. For this, we will enter into the matrix of X–Y data to be analyzed (abscissa in the first column ordered in the second):

data =

	0	1
0	1.00	2.00
1	2.00	2.60
2	3.00	23.16
3	4.00	27.57
4	5.00	24.26
5	6.00	16.63
6	8.00	28.50
7	11.00	47.20
8	12.00	50.03
9	13.00	60.33
10	14.00	59.89
11	16.00	71.18
12	17.00	84.27
13	19.00	77.69

This table represents an example of our data collected as a result of metrology

X= data

Y = data

n = lines (data)

Standard deviation: standard deviation

$X{:=}\ \text{data}^{\langle 0\rangle}$

$Y{:=}\ \text{data}^{\langle 1\rangle}$

n:= rows(data)

n:=14

Standard deviation

$$SD(x){:=}\ \text{stdev}(x)\cdot\sqrt{\frac{n}{n-1}}$$

Calculations of *characteristic statistics* according to the coordinates in X and in Y

Statistics	Coordinates in X	Coordinates in Y
Mean	mean (X) = 9.357	mean (Y) = 41.094
Median	median (X) = 9.500	median (Y) = 37.850
Variance	$SD(X)^2 = 35.786$	$SD(Y)^2 = 731.091$
Standard Deviation	SD(X) = 5.982	SD(Y) =27.039
	Regression Statistics	
Intersection at the origin	β_0 = ordon_orig (X, Y)	$\beta_0 = 8.042 \times 10^{-3}$
Gradient (*slope*)	β_1 = slope(X, Y)	$\beta_1 = 4.391$
Coefficient of correlation	corr (X, Y) =	0.9714
Coefficient R^2	corr $(X, Y)^2$ =	0.944
Covariance	covar(X, Y) =	145.905
Error type	Error type(X, Y) =	6.678
Graphs	$r(x) := \beta_0 + \beta_1 \cdot x$	$Scale = \max\left\|r(X)-Y\right\|.1.1$

Table 2.1. *Typical statistical characteristics for the case study*

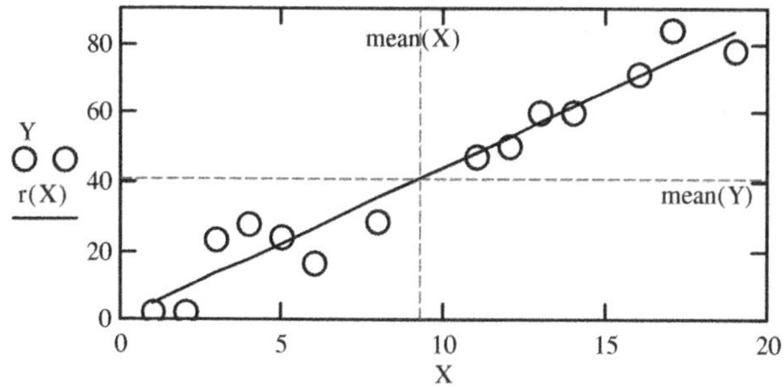

Figure 2.3. *Line of linear regression by the least-squares method*

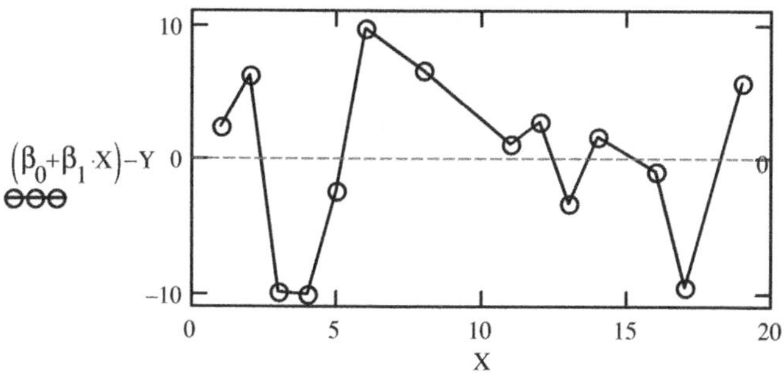

Figure 2.4. *Representation of the residues of linear regression*

To qualify the quality of adjustment for real data, based on the above trace (Figure 2.3), it is easy to see that the line on the base of the least squares of estimations of β_0 and β_1 constitutes an estimation of the true line of these simulated data (metrology data). The parameters in the functional part of the model (σ) are not generally known, but they could be estimated by looking at least-squares equations. The formula for the estimation is:

$$\hat{\sigma} = \sqrt{\frac{\Psi}{n-p}} = \sqrt{\frac{\sum_{i=1}^{n}\left(y_i - f\left(\vec{x}_i, \hat{\vec{\beta}}\right)\right)^2}{n-p}} \qquad [2.32]$$

(n) indicates the number of observations in the sample and (p) indicates the number of parameters in the functional part of model. $\hat{\sigma}$ is often indicated as the "residual standard deviation" of the process.

2.4.2. *Examples of nonlinear models*

These nonlinear models include:

$$\left\{\begin{array}{l} f\left(\vec{\tau};\vec{\beta}\right)=\dfrac{\beta_0+\beta_1\times\tau}{1+\beta_2\times\tau} \\ f\left(\vec{\tau};\vec{\beta}\right)=\beta_1\times\tau^{\beta_2} \\ f\left(\vec{\tau};\vec{\beta}\right)=\beta_0+\beta_1\times\exp^{-\tau\times\beta_2} \\ f\left(\vec{\tau};\vec{\beta}\right)=\beta_1\times\sin\left(\beta_2+\tau_1\times\beta_3\right)+\beta_4\times\cos\left(\beta_5+\tau_2\cdot\beta_6\right) \end{array}\right\} \quad [2.33]$$

a) Advantages of nonlinear least squares

Nonlinear regression, by the least-squares method, is widely used (see case study in Chapter 5, part 3 Volume 2). Although different mathematical processes can be described in this way using linear models, many processes exist, which are intrinsically nonlinear. Research on concrete resistance shows the asymptote approaches. Linear models do not describe asymptote processes well. Nonlinear regression can produce good estimations of unknown parameters in the model with relatively weak data. Another advantage resides in the prediction of calibration to respond to engineering issues.

In many cases, the interpretation of the interval that comes from nonlinear regression appears subject to caution. However, it is important to indicate the useful functioning of intervals. At the end of this chapter, we will present a case study, which relates to *interpolation by spline bidimensional function* (see Figure 2.30).

b) Disadvantages of nonlinear least-squares

The nonlinear regression models, which come from the least-squares method often go through iterative optimization procedures to calculate estimators. Thanks to linear functions in the parameters, the estimations of the least-squares parameters can always be obtained analytically, while this is generally not the case with nonlinear models. The use of iterative procedures of initial values for unknown parameters before the procedure of optimization cannot converge. Erroneous initial values can also cause convergence toward a local minimum rather than a global minimum, which defines least-squares estimations.

These disadvantages are highly sensitive to aberrant values. Just like a linear analysis of the least squares, the presence of one or two aberrant values in the data would seriously affect the results of a nonlinear analysis. Furthermore, it is unfortunate that there are fewer tools for validating the model for the detection of aberrant values in nonlinear regression than there are in linear regression.

There are three main parts to each model of the process, which are:

1) The response variable, usually written y;

2) The mathematical function, usually written: $f\left(\vec{x};\vec{\beta}\right)$;

3) Random errors, generally written:(ε).

The general form of the model is formulated: $y = f\left(\vec{x};\vec{\beta}\right) + \varepsilon$

The mathematical function is made up of two parts. These parts are predictive variables, x_1, x_2, ..., x_n and the parameters are: β_0, β_1, β_2, β_n,... The predictive variables are observed with a variable response. They are the quantities described as input locations in the mathematical function $f\left(\vec{x};\vec{\beta}\right)$. The collection of this group of predictive variables is written $\left(\vec{x}\right)$, that is, $\vec{x} = \left\{x_0, x_1, \ldots, x_n\right\}$

The parameters are the quantities that will be estimated during the modeling process. Their true values are unknown, apart from the simulation experiences. As for predictive variables, the collection of parameters is written $\left(\vec{\beta}\right)$, i.e.: $\vec{\beta} = \left\{\beta_0, \beta_1, \ldots, \beta_n\right\}$. The parameters and the predictive variables are combined in different forms to give the function used to describe deterministic variation in the response variable. For a straight line with an unknown interception and a slope, for example, there are two parameters and a predictive variable, which are:

$$f\left(x;\vec{\beta}\right) = \beta_0 + \beta_1 \cdot x \quad [2.34]$$

– for a *straight line* with a *known gradient* but an *unknown interception,* there would only be one parameter which would be: $f\left(x;\vec{\beta}\right) = \beta_0 + x$;

– for a *quadratic surface* with *two predictive variables,* there are six parameters which make up the complete model:

$$f\left(\vec{x};\vec{\beta}\right) = \beta_0 + \beta_1 x_1 + \beta_2 x_2 + \beta_{12} x_1 x_2 + \beta_{11} x_1^2 + \beta_{22} x_2^2 \quad [2.35]$$

***Random error*:** Like the parameters of the mathematical function, random errors are unknown. They are quite simply the difference between the data and the mathematical functions. They are thought to follow a particular probability distribution. The probability distribution which describes the errors has a mean of zero and an unknown standard deviation written (σ), which is another parameter in the model like β_s.

2.4.2.1. *Basic definition of the process modeling using examples*

Process modeling is the concise description of total variation within a single quantity y, breaking it down to a *deterministic component* given by a mathematical function of one or a few other quantities ($x_1, x_2, \ldots, x_n$), plus a random component which follows a particularly appropriate probability distribution. For example, the total variation of the roughness of a part may be described by the partition of the variable without a deterministic part, which works with the geometry of the tool (corner radius of a cup tool), as well as a few random errors such as those expressed below:

Attention: This case study was carried out in a production factory. We will deal primarily with the Weibull distribution to study convergence using the least-squares method (nonlinear case).

Roughness R$_a$- Case of resurfacing (Figure 2.5): For a_t = (0.05–16) feed rate per tooth and r = 1.6 mm corner radius of the tool.

$$R(a_t)=\left(\frac{20\times\sqrt{a_t}}{r^{0.12}}\right)\times 10^{-3} \qquad [2.36]$$

R(at) =

	0
0	$4.227\cdot 10^{-3}$
1	0.019
2	0.027
3	0.033
4	0.038
5	0.042
6	0.046
7	0.05
8	...

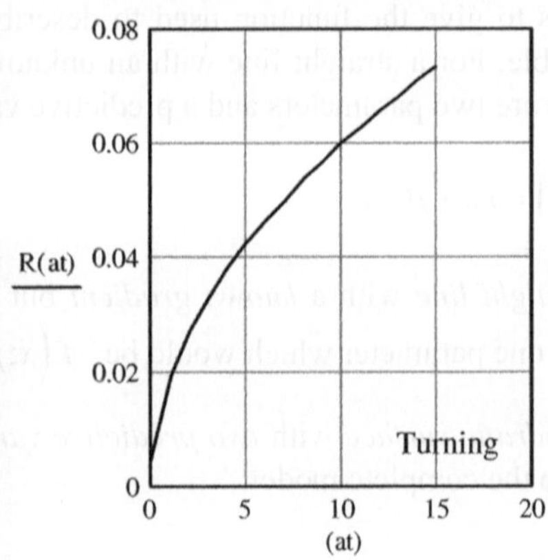

Figure 2.5. *Roughness evolution: resurfacing*

Roughness R$_a$- Case of conventional milling (Figure 2.6(a)): For a = 0.4 to 16; Z = 6 teeth and D = 70 mm, we will have:

$$R_a = R(a) = \frac{a^2}{8}\left(\frac{D}{2} + \frac{a \times Z}{\pi}\right) \times 10^3 \qquad [2.37]$$

Roughness R_A- Case of climb milling (Figure 2.6(b)): For the same data, $a = 0.4$ to 16; $Z = 6$ teeth and $D = 70$ mm, we will have:

$$R_A = R(A) = \frac{A^2}{8}\left(\frac{D}{2} - \frac{a \times Z}{\pi}\right) \times 10^3 \qquad [2.38]$$

Graphs and results of calculations:

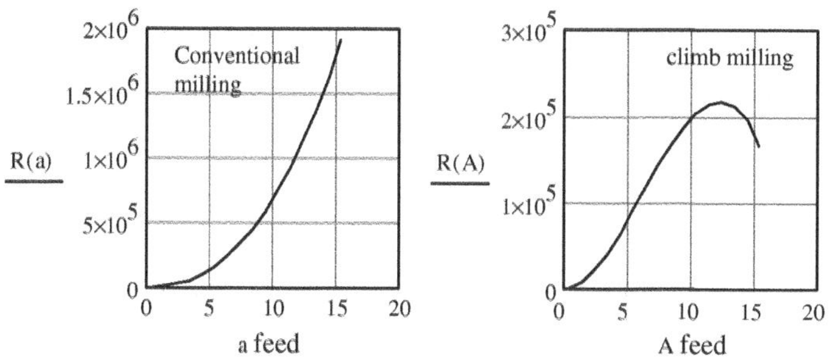

Figure 2.6(a). *Evolution of roughness: conventional milling*

We have manufactured a plaque in aluminum (SAE 6061), which has already been the object of a vast experimental campaign [GRO 09, GRO 11]. Below, we will present the results of one side, using an example.

	R_a, R_Z R_{max} → Roughness of zones from 1 to 10 → Plaque face in 6061: face A									
Ra	Ra_1	Ra_2	Ra_3	Ra_4	Ra_5	Ra_6	Ra_7	Ra_8	Ra_9	Ra_{10}
μm	0.62	0.62	0.62	0.62	0.45	0.40	0.56	0.55	0.65	0.40
μinch	24	24	24	24	18	15	22	22	26	16
Rz	Rz_1	Rz_2	Rz_3	Rz_4	Rz_5	Rz_6	Rz_7	Rz_8	Rz_9	Rz_{10}
μm	2.9	3.7	2.3	3	1.3	0.3	3.3	1	1.8	0.7
μinch	114	146	91	117	51	12	130	39	71	28
Rmax	Rm_1	Rm_2	Rm_3	Rm_4	Rm_5	Rm_6	Rm_7	Rm_8	Rm_9	Rm_{10}
μm	3.9	5.2	3.6	6.3	2.1	0.5	4	3.1	5.5	1.6
μinch	154	205	142	248	83	20	157	122	217	63

Table 2.2. *Experimental results of roughness of a 6061 rectified plaque*

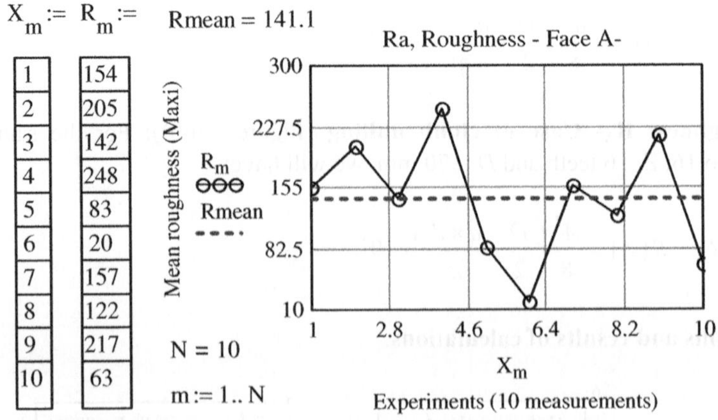

Figure 2.6(b). *Experimental results of roughness of a 6061 plaque*

Random errors cannot be individually characterized, but must be characterized according to a probability distribution, which describes the frequencies relative in occurrence of different sizes of error. Figure 2.5 shows the R_a data varying with the geometry of the tool in a deterministic way, except for a small quantity of random errors. The relationship between geometry and advancement is a curve, but it is not a perfect straight line. The Miner (convergence) method for the adjustment of the MC of nonlinear functions is presented in Chapter 2, Volume 2.

2.4.3. *Example of a more complex process*

The example of a straight line is simple in modeling. Another example is illustrated below.

The concept is identical to the straight line example, but the structure of the data is more complex. Variation (y) is dismantled into a deterministic part, which is contingent on another variable x, plus a certain more random variation.

The examples of physical process modeling have a single explanatory variable, but the concept easily extends to cases with more than one explanatory variable. The three-dimensional *parcels* below show an example with two explanatory variables. We have used the MathCAD program to model the physical process with two variables (ourselves). We have other cases of processes with two variables and with even more. Here is a concrete example from our own metrology and quality control laboratory:

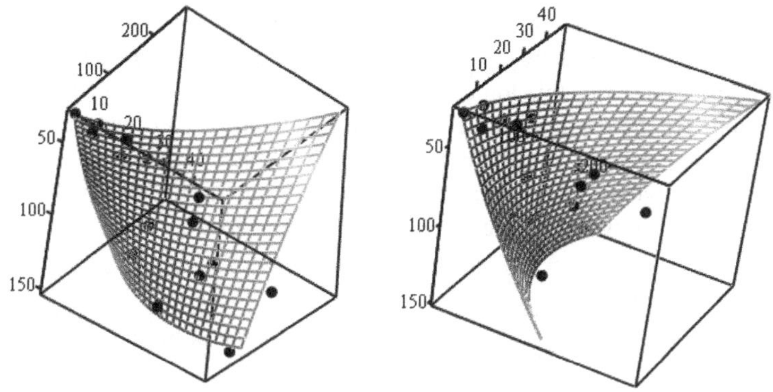

Figure 2.7. *Probability density curve, regression curve (cloud of 3D points) F,(X,Y,Z)= CreateMesh {f, min(X), max(X), min(Y), max(X), 25, 25}*

2.5. Conformity tests: adjustment and adequacy tests

Conformity tests allow us to judge the admissibility of a hypothesis concerning statistical distributions. The hypothesis is usually emitted with a certain probability (α) of not being admissible: If the real non-verification probability of the hypothesis is superior to (α), that implies that nothing is against accepting the hypothesis, without showing that it constitutes certain information. In the opposing case, it is better to reject the hypothesis for being too uncertain to translate into observed reality. We take for granted that a model used in the scientific field of engineering is well rehearsed and that the distribution law is completely *reserved* for it. This assumption is untrue. It is subject to caution and the results – however well calculated they may be – do not come from here and do not represent reality. The consequences of decision risk are often penalizing. The data are, in reality, supposed to come from a particular distribution.

Usually, we proceed to adequacy tests to control the *concordance* of a distribution law used for a model. The quality of the adjustment of these tests indicates whether it would be reasonable to suppose that a random sample will follow a distribution to represent a physical model. For example, we often bring forward the hypothesis that the Birnbaum–Saunders distribution represents material fatigue. *Is this always true*? *Of course not*!

In what name do we declare the dependability of these distributions to represent complex phenomena in terms of material behavior? It is well understood that it is in the name of experience and real life. This affirmation would not be able to be generalized in an industrial situation due to security risks (bridges, buildings, medicine, etc.) Statistical techniques are often based upon observations, which come from a population with a specific distribution. In quality control, test cards are the results of continuous measures. Therefore, testing demands that data come from a normal distribution. Does this logical idea to continue lifespan modeling in this way require a specific distribution model? We have noticed that in many cases, normal distribution is solicited in all situations. This shows a lack of scientific rigor. It is not the distribution, which is subject to caution, but rather the abusively erroneous usage, which is done with it. Here are some common statistical tests.

	Tests	Use(s)
Parametric	*T tests*	Tests on the mean(s) of one or two normal populations
	ANOVA	Test of equality of one or two means of normal populations
	Hotelling's T^2	Test of equality of two means from multi-normal populations
	Bartlett	Test of equality of variances from more than two normal populations
	Wilks	Test of equality of more than two means of multi-normal populations
	Dunnett	Test to compare groups with a group of reference
	Newman–Keuls	Test for multiple comparison
	Fisher–Irwin	Identity test of two Bernoulli populations

	Tests	Use(s)
NON-parametric	Chi square tests	Test of adequacy between a distribution and a sample
		Identity test of several distributions
		Independence test of two nominal variables
		Test of symmetry of a conjoined distribution
		Identity test of two marginal distributions
		Test of adequacy between a distribution and a sample
	Kolmogorov	Test of adequacy between a distribution and a sample
	Kolmogorov–Smirnov	Identity test for two distributions
	Mann–Whitney	Identity test for two distributions
	Kruskal–Wallis	Identity test for more than two independent populations
	Friedman	Identity test for more than two populations (matched samples)

Table 2.3. *Typical statistical tests in reliability and in quality control*

2.5.1. *Model of the hypothesis test for adequacy and adjustment*

An adequacy test tests:

– $\mathbf{H}_0$: Probability distribution, which generates the sample is $Y(\tau)$, against

– $\mathbf{H}_n$: Probability distribution, which generates the sample *is not* $Y(\tau)$.

Following the general construction schema of a test, we must therefore:

i) *First find a statistic (according to the sample), which intuitively supplies a correct indication of whether* $\boldsymbol{H}_0$ *is true or false; and*

ii) *Then calculate the probability distribution under* $\boldsymbol{H}_0$ *of this test statistic.*

Many tests have been analyzed in the literature, but only a few have been used in practice. We will now describe the most important ones. The adequacy tests are based on cumulative distribution function test:

$$\mathbf{H}_0 \text{ where } Y = Y_0 \text{ and } \mathbf{H}_n \text{ where } Y \neq Y_0$$

Y is the cumulative distribution function of the distribution, having generated the sample, and Y_0 is the cumulative distribution function of reference. We must identify a statistic, which is a *reasonable measurement* of the difference between the empirical cumulative distribution function and the true cumulative distribution function. Then, we will calculate its distribution under $\mathbf{H}_0$. The quality of hypothesis adjustment tests depends on two distinct hypotheses: H_0 being the null hypothesis and H_A being the alternative hypothesis, or put differently:

– $\boldsymbol{H}_0$: *The data follow a specific distribution (null hypothesis).*

– $\boldsymbol{H}_a$: *The data do not follow a specified distribution (alternative).*

The parameters can be considered or estimated by looking at the data. A statistical hypothesis is, in fact, said to be quantitative relative to the characteristics of a given population (size of fractures, lifespan of bearings, etc.). We know that the first characteristic is the mean (μ). The hypothetical mean of the process analyzed is represented by (μ_0). For reasons of "*cost*" and of time, we look to probabilistic approaches and to simulations based upon mathematical rigor. In this case, we consider a random variable of size (n) of the process, if this process is centered to (μ_0). If there is tendency toward displacement, we will bring the mean of sample ($\bar{x}$) into play.

– A simple hypothesis where the data of $\boldsymbol{H}_0$ come from a known normal distribution according to $N\ (\mu = 0,\ \sigma = 1)$.

– A hypothesis would be composite when the data of $\boldsymbol{H_0}$ come from an unknown normal distribution, that is, $N\ (\mu = ?\ \sigma = ?)$.

Composite hypotheses are more widespread because they allow us to decide whether a sample comes from a distribution of a specific type. The shape of the distribution lends itself to an independent interest in the value of parameters (μ, σ). Unfortunately, composite hypotheses are more difficult to put into practice, as the critical values are often difficult to calculate.

Problems of censored tests: In reliability, censored tests pose a few problems to tests. When data are censored, the values of the sample are somewhat limited. Censoring frequently arises in reliability tests for trial times where the number of failures to observe is fixed in advance. This point [MOO 86] will not be broached here. This is in spite of the fact that adequacy tests are numerous in the literature. Only three main types of tests [MON 91] will be covered in this text. Other specific criteria exist based on regression and graphical techniques. We will postulate a few of these based on case studies in Chapter 8, Volume 2.

Typical tests:

– Kolmogorov–Smirnov (KS) for continuous distributions,

– Anderson–Darling (AD^2) based on the cumulative distribution function (continuous), De Pearson, and Chi-square (χ^2) for continuos and discrete distributions.

2.5.2. *Kolmogorov–Smirnov Test (KS 1930 and 1936)*

Although severe, the KS test constitutes an excellent alternative to the χ^2 test. It is also true that it is only valid for testing continuous distributions, shown by the normal distribution of the Weibull distribution. The advantage of the χ^2 test compared to the KS test is that it is also used to test distributions, which are both continuous and discrete. It is for this reason that it will have special status in the treatment of our data, in this text. Additionally, the KS test (see Appendix Table 2) is one of numerous tests, which are based on the empirical cumulative distribution function.

The critical values associated with the statistics of test D, are difficult to calculate for the size of finite samples [BIR 52]. It is sometimes necessary to go through the Monte Carlo distribution. The KS [CHA 67] test is used to decide if the sample comes from a population having a specific distribution. Considering *N-ordered* points of data Y_1, Y_2, Y_3, ..., Y_N, the cumulative distribution function is defined as follows:

$$E_N = \{n(i)/N\} \tag{2.39}$$

where ***n(i)*** is the number of points less than $\boldsymbol{Y_i}$, which are the points increasing in order. It is a step function, which increases from (1/*N*) to the value of each ordered point of data. The graph below is a representation of the empirical cumulative distribution function with a normal cumulative distribution function for 100 random values from a normal distribution. The KS test is based on the maximal *max(D)* distance between these two curves.

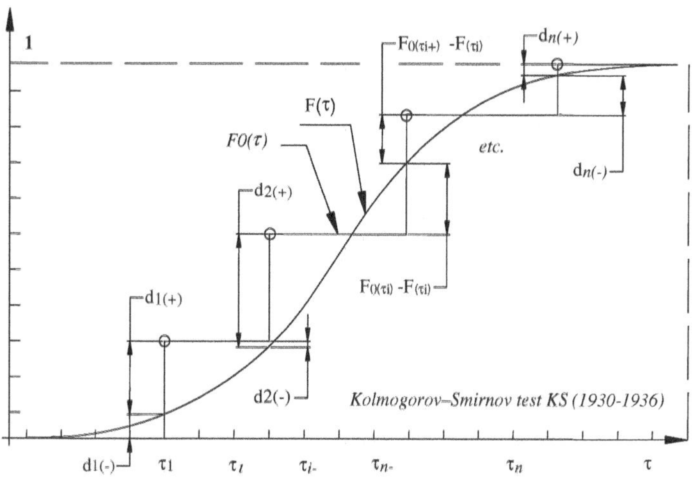

Figure 2.8. *Graph showing the KS test*

The KS test applied on a model of a sample of an unknown distribution ($\wp$) is an adjustment test dedicated to the ensemble of quantiles. The null hypothesis is represented by H_0. Distribution ($\wp$) has Y_0 for its centered, reduced, and continuous cumulative distribution function. This function would therefore have $\hat{Y}_0$ as the empirical function of cumulative distribution, from the closest sample possible of Y_0 of R in [0, 1] which is expressed as:

$$\hat{Y}(\tau) = \begin{Bmatrix} 0 \quad for \ \tau \prec \tau_1 \\ \left(\dfrac{i}{n}\right) \quad for \ \tau_i \leq \tau \prec \tau_{(i+1)} \\ 1 \quad for \ \tau \geq \tau_n \end{Bmatrix} \tag{2.40}$$

τ_i are the values of the sample arranged in ascending order. In other words, $\hat{Y}(\tau)$ is the proportion of elements of the sample, which are inferior or equal to (τ). We measure the **adequacy** of the empirical cumulative distribution function to function Y_0 by *max*(D), a distance known to be from KS. It is, in fact, the distance of the uniform norm between two cumulative distribution functions (see Figure 2.8). To calculate this distance, it suffices to evaluate the difference between $\hat{Y}(\tau)$ and Y_0 to the points τ_i:

$$D_{KS}\left(Y_0,\hat{Y}(\tau)\right)=\max_{i=1\to n}\left\{\left|Y_0(\tau_i)-\left(\frac{i}{n}\right)\right|,\left|Y_0(\tau_i)-\left(\frac{i-1}{n}\right)\right|\right\} \qquad [2.41]$$

In the case of the H_0 hypothesis, the statistical distribution $D_{KS}\left(F_0,\hat{F}(\tau)\right)$ does not depend on Y_0, as the images of τ_i by F_0 are the RV of U distribution [0, 1]. The cumulative distribution function of $D_{KS}\left(F_0,\hat{F}(\tau)\right)$ does not have a simple, explicit expression; it must be numerically calculated. For the samples of sufficient size, that is, the hypothesis H_0, we have, $t \geq 0$ for everything and we look to the asymptotic result:

$$\lim_{n\to\infty}\wp_{H_0}\left\{\sqrt{n}\times D_{KS}\left(Y_0,\hat{Y}(\tau)\right)\leq t\right\}=1-2\sum_{k=1}^{+\infty}(-1)^{k+1}\times Exp^{-2k^2t^2} \qquad [2.42]$$

The series converges very rapidly. In practice, it is for $t > (\mathbf{0.56})$ that the sum of the three first terms supplies an approximation with an error inferior to 10^{-4}. In the case where hypothesis H_0 would be false, that is, when $\sqrt{n}\times D_{KS}\left(F_0,\hat{F}(\tau)\right)$ leans toward ∞ with ($n > 0$), the test becomes, in reality, unilateral to the right and we would then reject the values that are too high. In the case where the distance $D_{KS}\left(F_0,\hat{F}(\tau)\right)$ would take the value **0.047** for a sample of size n = 1,000; the statistic $\sqrt{n}\times D_{KS}\left(Y_0,\hat{Y}(\tau)\right)$ is worth $t = 1.486$. ($\wp$-value) thus corresponding to:

$$p(t)\simeq 2\sum_{k=1}^{3}(-1)^{k+1}\times Exp^{-2k^2t^2}=0.0241$$

An interesting characteristic of the KS test is that the distribution of the statistic itself does not depend on the cumulative distribution function, which underlies the trial. It is a very accurate test (the χ^2 of validity of the adjustment of the test relies upon the sample having an adequate size for the approximations to be valid). In spite of these advantages, the KS test has other limits, which penalize it:

– It applies uniquely to continuous distributions.

– It shows obvious sensitivity close to the center of distribution and to the line.

– If the parameters of localization (λ), of scale (η) and of shape (β) are estimated based on the data, the critical region of the KS test is no longer valid. It is therefore necessary to proceed to a Monte Carlo simulation.

n Sample size	Critical thresholds $\mathbf{D}_\alpha\mathbf{(n)}$				
	α = **.20**	α = **.15**	α = **.10**	α = **.05**	α = **.01**
1	.900	.925	.950	.975	.995
2	.684	.726	.776	.842	.929
3	.565	.597	.642	.708	.828
4	.494	.525	.564	.624	.733
5	.446	.474	.510	.565	.669
6	.410	.436	.470	.521	.618
7	.381	.405	.438	.486	.577
8	.358	.381	.411	.457	.543
9	.339	.360	.388	.432	.514
10	.322	.342	.368	.410	.490
11	.307	.326	.352	.391	.468
12	.295	.313	.338	.375	.450
13	.284	.302	.325	.361	.433
14	.274	.292	.314	.349	.418
15	.266	.283	.304	.338	.404
16	.258	.274	.295	.328	.392
17	.250	.266	.286	.318	.381
18	.244	.259	.278	.309	.371
19	.237	.252	.272	.301	.363
20	.231	.246	.264	.294	.356
25	.210	.220	.240	.270	.320
30	.190	.200	.220	.240	.290
35	.180	.190	.210	.230	.270
>35	$1.07/\sqrt{n}$	$1.14/\sqrt{n}$	$1.22/\sqrt{n}$	$1.36/\sqrt{n}$	$1.63/\sqrt{n}$

Table 2.4. *Tabulated values of KS of the critical thresholds $D_\alpha(n)$ (see appendix)*

Due to the two final limitations evoked, many analysts prefer the Anderson–Darling tests to the KS test. However, the Anderson–Darling test is only available for use on a few specific distributions. We will not go deep into the workings of this test in this chapter. The Kolmogorow–Smimov test is defined by:

– H_0 The data follow a specified distribution.

– H_n The data do not follow a specified distribution.

The statistics of the Kolmogorov–Smirnov test are defined as:

$$D = \max_{1 \le i \le N} i \left\{ \left(F(Y_i) - \frac{i-1}{N} \right), \left(\frac{i}{N} - F(Y_i) \right) \right\} \qquad [2.43]$$

$F(Y_i)$ is the cumulated theoretical distribution of the tested distribution, which must be a continuous distribution (i.e. not discrete distributions like binomial distribution or the Poisson distribution), and must be entirely specified (i.e. parameters of localization (λ), of scale (η) and of shape (β) cannot be estimated based on data). This formula clearly indicates that the superior limit on the difference between these two formulae is (i/N). For real data, the difference is susceptible to being inferior to the superior limit. For example, for n = 20, the limit greater than the difference between these two formulae is from 0.05 (the critical value is from 5% to 0.294%).

2.5.2.1. *Example of the KS test for Y_1*

H_0: null hypothesis where the distribution corresponds to the data;

H_n: hypothesis where the distribution does not adapt to the data.

Data	Relative frequency	Observed frequency	Expected frequency	Absolute Difference
0.10	0.3333	0.3333	0.0952	0.2382
0.30	0.3333	0.6667	0.2592	0.4075
0.40	0.3333	1.0000	0.3297	0.6703
Results	Maximum of the absolute difference = **0.6703** The *p-value* is (0.15) and (0.1) → The data could be follow a Weibull distribution with a shape parameter of β= 1 and a scale parameter of η = 1			

Table 2.5. *Maximum of the absolute difference*

2.5.3. *Simulated test (1st application)*

Data value	Relative frequency	Observed frequency	Expected frequency	Absolute Difference
0.10	0.3333	0.3333	0.5481	0.2148
0.30	0.3333	0.6667	0.5879	0.0787
0.40	0.3333	1.0000	0.5985	0.4015

Table 2.6. *KS test of maximum difference*

– Maximum absolute difference = **0.4015**;

– The value of $p-n$ is > **(0.2)** → so, H_0 is accepted.

The data could *probably* follow a Weibull distribution with a shape parameter of $\boldsymbol{\beta}$= **0.1** and a scale parameter of $\boldsymbol{\eta}$ = **1**.

2.5.4. *Simulated test (2nd application)*

Data value	Relative frequency	Observed frequency	Expected frequency	Absolute Difference
0.10	0.3333	0.3333	0.2326	0.1007
0.30	0.3333	0.6667	0.2559	0.4108
0.40	0.3333	1.0000	0.2622	0.7378

Table 2.7. *KS test of maximum difference*

– Maximum absolute difference = 0.7378;

– The value of $p-n$ is(**0.1**) and (**0.05**).

Many procedures and other tests are based on the specific cumulative distribution of hypotheses, one of which, is normality. In fatigue reliability, experiments have shown that many lifespans follow a Birnbaum–Saunders or a Weibull distribution. Many non-parametric and robust techniques exist, which are not based on strong distribution hypotheses. To understand robustness, we must first understand a statistical technique, which works well on a large range of cumulative distribution function hypotheses. However, techniques based on specific hypotheses of cumulative distribution are, in general, more powerful than non-parametric techniques and *presumed* robust. That is understood due to the capacity to detect a difference, however minimal, when it exists in reality. Consequently, if the cumulative distribution hypotheses can be confirmed, parametric techniques have a higher status and allow us to obtain reliable results concerning decisions, which would be interdependent with them.

2.5.5. *Example 1*

Here is a series ordered from $n = 10$ pseudorandom numbers of experimental value represented by the table of experimental data:

$\tau =$

	0
0	8.430000
1	8.700000
2	9.330000
3	10.000100
4	11.000000
5	13.336670
6	15.021210
7	16.000000
8	16.117117
9	17.000000

Statistics of values $i := 0..9$ and $1 \le i \le n$

$y := (\tau)^{\langle 0 \rangle}$ $n := \text{rows}(\tau)$ $n = 10$

Number of points (Experiments) $n = \text{length}(\tau)$ $n = 10$

Sample mean: $\mu e := \text{mean}(\tau)$ $\mu e = 12.494$

Standard deviation $\sigma e := \text{stdev}(\tau)\cdot\sqrt{\dfrac{n}{n-1}}$ $\sigma e = 3.37$

Variance $VAe := \sigma e^2$ $VAe = 11.355$

First, we must calculate the cumulative distribution function for each experimental and theoretical piece of data and then work out their respective differences. We notice that when n increases, $Y_n(\tau)$ gets closer to $Y(\tau)$. The KS test compares the cumulative distribution function [$(pnorm(\tau, \mu, \sigma)$] with the empirical cumulative distribution function of the sample (experiments). The idea is simple. It consists of calculating the maximal distance *max*(D) (see Figure 2.8) between the theoretical and empirical function. If the maximal difference exceeds a certain value, which we can see from Table 2.4, we will say that the sample is an unsuitable one, or that it has been rejected subsequent to a KS test. Here is the reasoning:

1) Generate n random numbers between [0 and 1] and sort them in ascending order: $T_1, T_2, \ldots, T_n$.

2) Calculate distance D with the help of:

$$\left\{ D^{\sup} = \max_{1 \le i \le n} \left\{ \frac{i}{n} - \tau_i \right\} \; and \; D^{\inf} = \max_{1 \le i \le n} \left\{ \tau_i - \frac{i-1}{n} \right\} \right\}$$

3) Determine the critical value $D_\alpha(n)$ in Table 2.4.

4) If $\boldsymbol{D} > \boldsymbol{D}_{(\alpha,\, n)}$, we say that the *sample is unsuitable* (see Table 2.4).

We find D as explained in the second listed point above. Following this, we will pursue this reasoning:

The cumulative distribution function *pnorm (data,* μ_e*,* σ_e*)* is calculated based on experimental data: mean and standard deviation of a distribution that was unknown until now.

For that, we must calculate the cumulative distribution function for each piece of experimental and theoretical data. Then, we must calculate their respective differences, i.e.: Cumulative distribution function: **"pnorm(data,μe,se)** calculated based on experimental data, with t mean and standard deviation of a distribution, unknown until now:

pnorm(τ, μe, σe) =

	0
0	0.113928
1	0.130130
2	0.173913
3	0.229664
4	0.328803
5	0.598791
6	0.773412
7	0.850969
8	0.858892
9	0.909448

We will now calculate the cumulative distribution functions of the 10 pieces of data from (τ_i).

Y_i = *Y (cumulative distribution);* Y_{ic} = Y_i *(cumulative distribution),* and we will now calculate the differences between the cumulative distribution functions and the experimental and theoretical data. Then, we will calculate the probability functions, cumulated out of 10 pieces of data (τ_i).

Yi = Cumulative function and Yic=FiCumulative function

	Yi =	Yic =	LowerDeviation =	UpperDeviation =
	0	0	0	0
0	0.000	0.100	0.113928	-0.013928
1	0.100	0.200	0.030130	0.069870
2	0.200	0.300	-0.026087	0.126087
3	0.300	0.400	-0.070336	0.170336
4	0.400	0.500	-0.071197	0.171197 ↔
5	0.500	0.600	0.098791	$1.208844\cdot10^{-3}$
6	0.600	0.700	0.173412	-0.073412
7	0.700	0.800	0.150969	-0.050969
8	0.800	0.900	0.058892	0.041108
9	0.900	1.000	$9.447895\cdot10^{-3}$	0.090552

D is superior distance

LowerDeviation $\equiv$ pnorm(τ, μe, σe) – Yi and

UpperDeviation $\equiv$ Yic – pnorm(τ, μe, σe) So max(UpperDeviation) = 0.171197

2.5.5.1. *Graphs and comments*

The calculation has allowed us to find the maximal distance in the Kolmogorov–Smirnov direction, that is, max(D) = max (upper deviation) = **0.171197**. By reading the KS table, we have been able to compare our result with $D_{(\alpha,\, n)} = D\,(5\%, 10)$ = **0.410.** Based on this result, we deduce that $D_{\text{calculated}} < D_{(\alpha,\, n)}$, that is, 0.171197 < 0.410, that the KS test accepts the hypothesis of this sample, according to which empirical law follows the theoretical distribution laws of a normal distribution (Gaussian of mean, μe = 12.494 and of standard deviation σ_e = 3.37). Here is a graph showing our results:

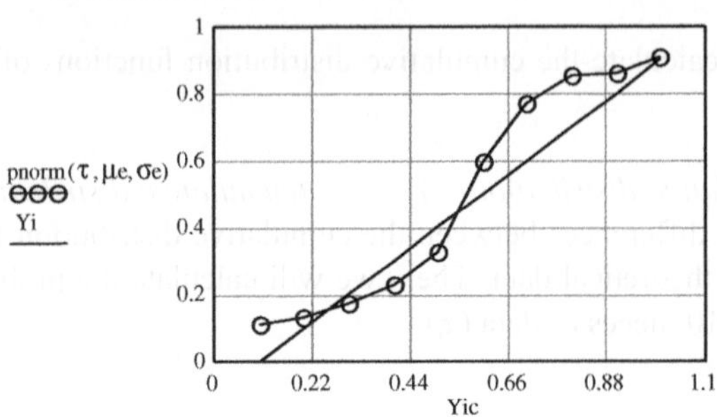

Figure 2.9. *Kolmogorov–Smirnov (KS) Test*

Here is the graph showing two distributions, which are presumed normal:

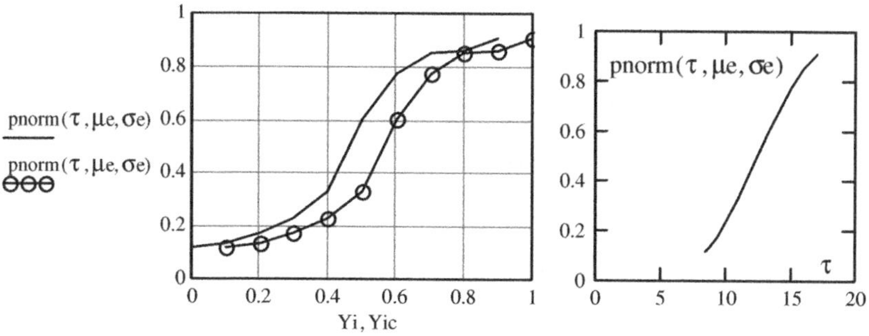

Figure 2.10. *Kolmogorov–Smirnov (KS) Test; two distributions*

We have noticed that in cases of small samples, the KS adequacy test is *inefficient*. For example, at the same threshold of 5% of the KS table, for the increased sample size ($n \geq 35$) the KS test would have been invalid on inadequacy. Here is the proof: $D_{(\alpha,\ n)} = D\ (5\%, 35) = 0.239995$.

According to sample size (n), the test *oscillates* from being adequate to being inadequate. With ($n \geq 35$), looking at the KS table shows $D_{(\alpha,\ n)} = D(5\%, 35) = 0.239995$. Based on this, we conclude that in this case, the hypothesis will have been restricted at the edge, correct to $n > 36$, we will have $\left(136/\sqrt{n}\right) = 22.667...$, which would lead us to reject the sample hypothesis and obtain more information on adequacy.

2.5.6. *Example 2 (Weibull or not?)*

We will proceed using the same methodological approach to try and see whether the experiment shows a Weibull distribution with two parameters applied to the lifespan distributions of materials and structures.

2.5.6.1. *Chance draw of 15 random numbers (unknown distribution)*

We have taken 10 values at random (we could have taken 100, 1,000, or more, the pedagogical approach would be the same) $\beta = 2.5$ (shape parameter) and $m = 10$. We have used the previously programmed function ***r****Weibull (β, m)*.

rweibull(m, β) =

	0
0	1.220
1	0.779
2	1.019
3	0.520
4	1.250
5	0.651
6	1.072
7	1.418
8	1.297
9	0.168

After sorting in ascending order

$$\tau := \begin{pmatrix} 0.481 \\ 0.583 \\ 0.602 \\ 0.807 \\ 0.825 \\ 0.862 \\ 0.915 \\ 1.168 \\ 1.19 \\ 1.28 \end{pmatrix}$$

Statistics on the data: Number of points $n = 10$, mean $\mu = 0.871$, and standard deviation $\sigma = 0.373$. We must find the maximum distance max (D) in the KS direction, as before:

pweibull(τ, β) =

	0
0	0.148247
1	0.228578
2	0.245111
3	0.442915
4	0.461091
5	0.498359
6	0.551054
7	0.771078
8	0.786642
9	0.843334

For β = 2.5 and η := 5

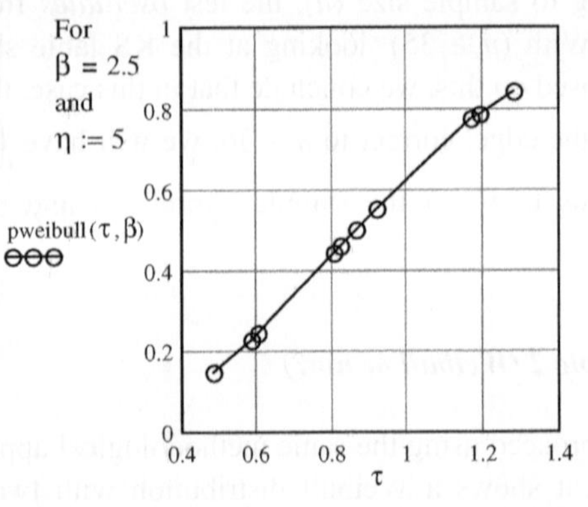

Figure 2.11. *Kolmogorov–Smirnov (KS) Test; Weibull distribution*

We will now calculate the cumulative distribution functions using 10 pieces of data, that is, $F_i = F_{\text{cumulative dist}}$ and $F_{ic} = F$ cumulative distribution (cumulative probability). Then, we will calculate the difference between the cumulative distribution functions of the experimental and theoretical data (simulated):

$$LowerDeviation = pnorm\{\tau, \mu_{Exp}, \sigma_{Exp}\} - F_i \text{ and } UpperDeviation = F_{ic} - pnorm\{\tau, \mu_{Exp}, \sigma_{Exp}\}$$

	Fi =	Fic =	LowerDeviation =	UpperDeviation =
	0	0	0	0
0	0.00	0.10	0.076730	0.023270
1	0.10	0.20	0.045855	0.054145
2	0.20	0.30	-0.037661	0.137661 ↔ *D superior distance*
3	0.30	0.40	0.107043	$-7.042561 \cdot 10^{-3}$
4	0.40	0.50	0.032769	0.067231
5	0.50	0.60	-0.013566	0.113566
6	0.60	0.70	-0.036511	0.136511
7	0.70	0.80	0.161061	-0.061061
8	0.80	0.90	0.078103	0.021897
9	0.90	1.00	0.032504	0.067496

2.5.6.2. *Graphs, conclusion, and comments*

This calculation has allowed us to find the maximal distance in the Kolmogorov–Smirnov direction, i.e. *max*(*D*) = max (upper deviation) = **0.137661**. Reading the KS table has allowed us to compare our results with $D_{(\alpha,\ n)} = D$ (5%, 10) = **0.410.** We deduce, based on this result $D_{\text{calculated}} < D_{(\alpha,\ n)}$, i.e. 0.137661 < 0.410, that the KS test accepts the hypothesis of this sample, according to which empirical law follows theoretical distributions of the *Weibull* distribution with two parameters of mean $\mu_e = 0.871$ and of standard deviation $\sigma_e = 0.273$.

Here is the graph showing our results, relative to distance *D*:

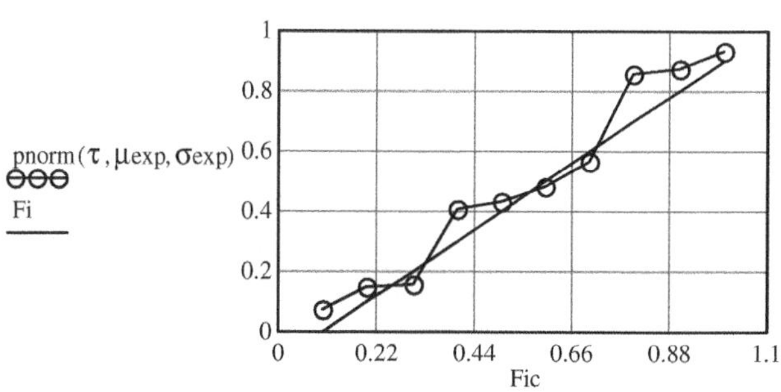

Figure 2.12. *Curve of the distance of the Weibull law with two parameters according to the KS test*

Here is the graph showing the two Weibull distributions.

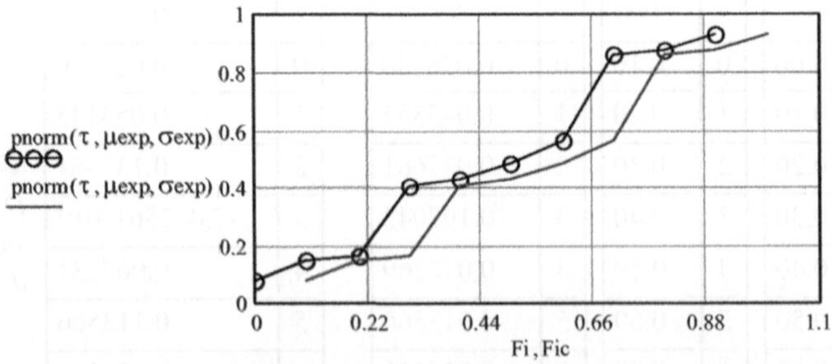

Figure 2.13. *Weibull cumulative distribution curves (theory/practice) with two parameters*

Distribution and cumulative distribution functions of the Weibull distribution:

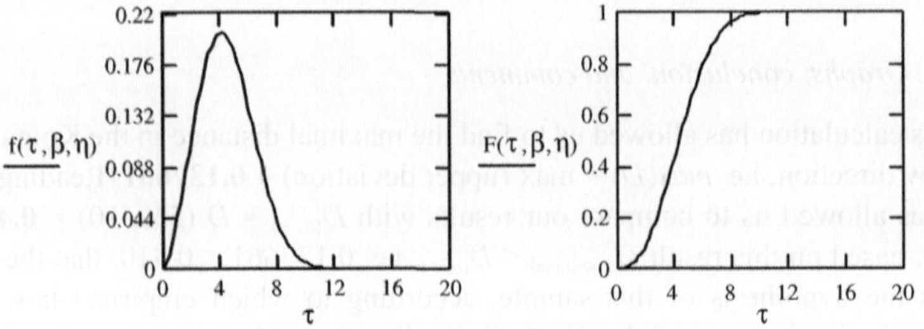

Figure 2.14. *Probability density and cumulative distribution curves of the Weibull distribution with two parameters*

Inverse cumulative distribution functions and rate of failure of the Weibull distribution:

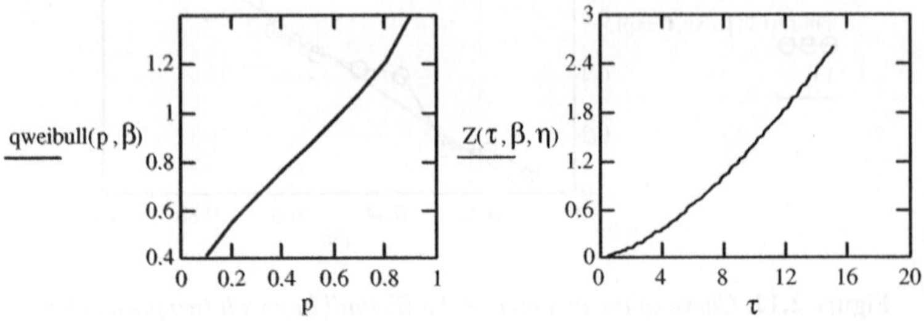

Figure 2.15. *Inverse cumulative distribution curves of the Weibull distribution with two parameters*

Reliability function $R(\tau)$ and number of random variables taken by chance, which allow us to follow the Weibull distribution with two parameters.

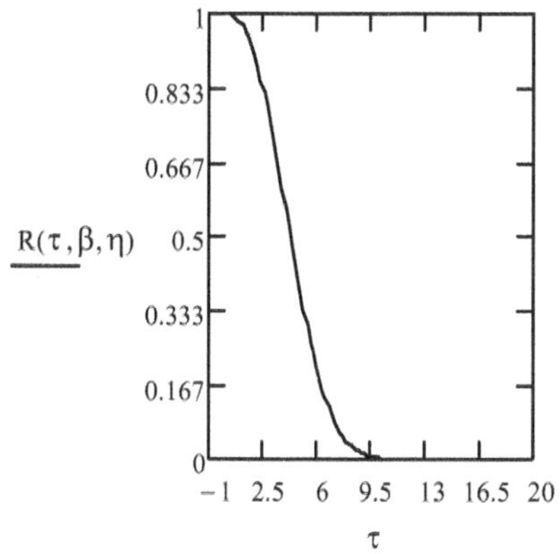

rweibull(m, β) =

	0
0	0.902
1	0.466
2	0.573
3	0.101
4	0.753
5	1.119
6	0.497
7	0.991
8	0.686
9	1.862

rweibull(m, β) =

	0
0	0.776
1	0.5
2	0.879
3	0.615
4	0.906
5	0.614
6	0.765
7	0.624
8	0.792
9	1.289

etc. ...

Figure 2.16. *Reliability curve of the Weibull distribution with two parameters*

2.5.7. *Cramer–Von Mises (CVM) test*

The KS test is always conceived based on a single observation of the sample and a single point from the cumulative distribution function contestant. It would be beneficial to measure the difference between the two cumulative distribution functions by comparing these functions based on the integral of their domain [$-\infty$ to $+\infty$]. It remains true that many tests exist whose statistics are based on the integral of the square of the difference $[Y_n(\tau)-Y(\tau)]^2$ between the empirical cumulative distribution function and the reference cumulative distribution function. The simplest aspect of this statistic is:

$$\delta = \int_{-\infty}^{+\infty} \left[Y_n(\tau) - Y_n(\tau)\right]^2 d\tau \qquad [2.44]$$

We notice that this defines itself simply as the area between the empirical cumulative distribution function and the reference cumulative distribution function. This statistic is not used, as the calculation of it in its entirety is unworkable, in general. Therefore, we introduced an arbitrary factor to be considered $d\kappa(x)$:

$$\delta = \int_{-\infty}^{+\infty} \left[Y_n(\tau) - Y_n(\tau)\right]^2 d\kappa(\tau) \qquad [2.45]$$

If δ is chosen carefully, it can allow the calculation of the integral. The most common choice for $\kappa(\tau)$ is $Y(\tau)$ itself. The statistic ω^2, which results from this addition, is named the Cramer–Von Mises (CVM) statistic, which is defined by:

$$\omega^2 = n \times \int_{-\infty}^{+\infty} \left[Y_n(\tau) - Y_n(\tau) \right]^2 dY(\tau) \qquad [2.46]$$

The essential advantage of this choice of $\kappa(\tau)$ is that the integral can now be calculated as a sum only bringing the values of $Y(\tau)$ into play for observations. $y_{(i)}$l indicates the statistic of order in row i of the sample. We can show that:

$$\omega^2 = \sum_{i=1}^{n} \left(Y(\tau_i) - \frac{2i-1}{2n} \right)^2 + \left(\frac{1}{12n} \right) \qquad [2.47]$$

The asymptotic function of cumulative distribution of ω^2 is known, but is very complex to establish. This text deals with reliability, with quality control, and with uncertainty. Other texts, which are more specialized in advanced statistics, propose solutions to the problems that precede this. The reader would undoubtedly benefit from consulting these.

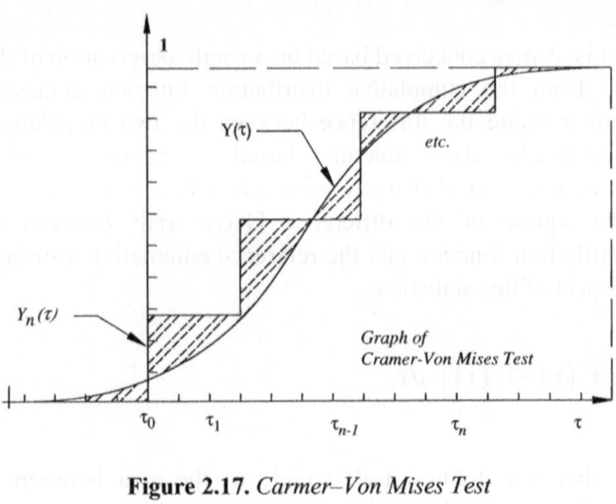

Figure 2.17. *Carmer–Von Mises Test*

2.5.8. *The Anderson–Darling test*

The Anderson–Darling test [STE 74, STE 76, STE 77, STE 79] is used to test whether a sample of data comes from a specific distribution. It is a matter, which relates to the modification of the KS test which puts more emphasis on the distribution line than the KS does. The KS test is a free distribution in the sense that the critical values do not depend on the specific distribution of the trial. The Anderson–Darling test has the advantage of allowing a more responsive trial. Its disadvantage resides in the essential values having to be calculated for each distribution, as this test is applied with the help of programs, which are appropriate for supplying relevant critical values. The Anderson–Darling test is defined by:

– H_0: *Data which follow a specified distribution*;

– H_n: *Data which do not follow a specified distribution.*

Below is the Anderson–Darling statistic:

$$A^2 = -(N+S) \text{ where } S = \sum_{i=1}^{N} \frac{(2i-1)}{N} \times \left\{ LnF(Y_i) + Ln\left(1 - F\left(Y_{N+1-i}\right)\right) \right\} \qquad [2.48]$$

F(.) is the cumulative distribution function of the specified distribution. We must clearly state that the Y_i are the data *ordered* by the Anderson–Darling test. The Anderson tests give the advantage to the normality test on the Y_I function.

1) Anderson–Darling: **SAMPLE 1**

The data come from a normal distribution (Laplace–Gauss)

2) Statistics:

- Number of observations = **1,000**

- Mean (μ) = **0.4359940E-02**

- Standard deviation (s or σ)= **1.001816**

- Statistical value of the Anderson–Darling TEST = 0.2565918

- Adjusted statistical value of the test = 0.2576117

3) Essential values:

- to 90% = 0.6560000 Point

- to 95% = 0.7870000 Point

- to 97.5% = 0.9180000 Point

- to 99% = 1.092000 Points

At the **5%** threshold, the data do not come from a normal distribution.

The Anderson–Darling test consists of comparing the theoretical distribution $Y_0(\tau)$ with the experimental distribution $Y(\tau)$ while calculating the statistic below:

$$AD^2 = \int_{-\infty}^{+\infty} \left(Y_n(\tau) - Y(\tau)\right)^2 \times \omega(\tau)\, dY(\tau) \qquad [2.49]$$

where $\omega(\tau)$ is a weight function. The standard Anderson–Darling case corresponds to the weight function below:

$$\omega(\tau) = \frac{1}{Y_0(\tau)\times\left(1 - Y_0(\tau)\right)} \qquad [2.50]$$

This formula allows us to give more influence to the weak and strong frequencies. This leads to the statistic marked AD².

dbeta(τ, p1, q1) =

0.000
3.199
2.274
1.866
1.624
1.461
1.340
1.248
1.173
...

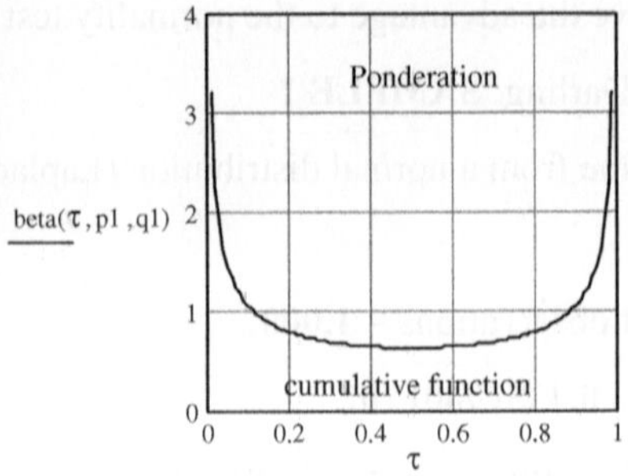

Figure 2.18. *Graph showing the distribution of cumulative probabilities*

The *weighting* function of the Anderson–Darling (AD²) test by modifying the weighting function in $\omega(\tau) = 1/\left(1 - Y_0(\tau)\right)$, we obtain a test that is sensitive to the behavior of rare frequencies. This testing procedure can prove to be particularly useful when looking at extreme values. A notable weakness of the CVM test comes from the difference between the empirical cumulative distribution function and the reference cumulative distribution function leaning toward 0 when $\tau \to -\infty$ or $\tau \to +\infty$. The value of ω^2 is therefore not very sensitive to exact positions of "*far away*" observations in the distribution lines. This is unfortunate. Often, it is a question of knowing whether the distribution of rare events (represented by these far away observations) *previously* conform to a certain idea about the nature of global distribution. A modification of the elementary statistic L_2 consists of introducing a weighting function, which gives back their importance to far away observations. The *weighting* function which is most commonly used is:

$$\omega(\tau)=\frac{1}{Y(\tau)\times\left[1-Y(\tau)\right]} \quad [2.51]$$

This expression is minimal in comparison to the surroundings of the median. It leans toward the infinite when τ leans toward $-\infty$ or $+\infty$. In fact, we can show that $Y(\tau) \times [1 - Y(\tau)]$ is the variance of the empirical cumulative distribution function in τ, to the effect that the test statistic is now integral to the square of the standardized difference between the empirical cumulative distribution function and the reference cumulative distribution function. The statistic AD^2 is:

$$AD^2 = n\times\int_{-\infty}^{+\infty}\left(Y_n(\tau)-Y(\tau)\right)^2\times\left(\frac{1}{Y(\tau)-1-Y(\tau)}\right)d(\tau) \quad [2.52]$$

We normally call (AD^2) the Anderson–Darling statistic. This can also be expressed in the form of a sum formulated below:

$$AD^2 = -\left(n+\frac{1}{n}\right)\times\sum_{i=1}^{n}(2i-1)\times\left\{ LogY(\tau_i)+LogY\left[1-Y(\tau_{n+1-i})\right] \right\} \quad [2.53]$$

All of the adequacy tests based on the comparison between the "empirical cumulative distribution function" and the "*candidate* cumulative distribution function" have a certain number of properties in common. It is the case where the distribution does not depend upon the candidate distribution. For example, let us consider the Kolmogorov statistic. Its distribution depends on sample size (n), but does not depend on the distribution tested when this is completely determined. The same applies to the CVM statistics and the Anderson–Darling statistics. This can be explained by the fact that the transformation by cumulative distribution function (TCDF) transforms independent observations which come from all probability distributions into independent uniform variables distributed in [0, 1].

The considerations, which relate to the behavior of the test statistic transform into considerations which relate to the properties of sampling of the unique distribution $U[0, 1]$. If this informal argument is judged to be unconvincing, we would make it explicit by demonstrating it within Kolmogorov's statistical case. The demonstration adapts without difficulty to any kind of statistic based on the empirical cumulative distribution function. This is no longer true when the distribution tested is **not** completely determined, and some of its parameters must be estimated based on the sample.

Estimated parameters: One of the *good properties* of the χ^2 test is the fact that it has well-understood effects to estimate the parameter values of the distribution of test statistics. For tests based on empirical cumulative distribution functions, things are less simple. In particular, it is no longer true that the distribution of this statistic does not depend on the considered distribution and it is necessary to proceed to tabulations (using the Monte Carlo simulation) for numerous distributions (normal, exponential, gamma, Weibull, logistic, etc.), as well as for different sample sizes. The test procedure must therefore begin by proceeding with the vector estimation (θ) of unknown parameters and using the corresponding cumulative distribution function $Y(\tau, \theta)$ as a reference cumulative distribution function to calculate the value of the test statistic.

It is often possible to add corrective terms (which do not depend on sample size) to standard statistics in such a way that the *corrected* statistical value can be used with the unique table of critical values of the *asymptotic* distribution. However, we can show that if the estimated parameters are localization parameters (γ) and/or scale parameters (η), the distribution of the test statistics (although dependent on the nature of the distribution) does *not* depend on the value of the(se) parameter(s). Demonstrating this result is difficult, but we can present a heuristic argument in its favor. Although the reference cumulative distribution function depends on the sample (due to parameter estimation), ideally the distribution of the test statistics does not depend on the parameter values of localization and scale.

Two independent samples: *All of these adequacy tests, based on the empirical cumulative distribution function, can be transformed into tests, which broach the question of whether or not two independent samples come from identical distributions.*

Discrete distributions: *We have attempted to adapt all adequacy tests based on the empirical cumulative distribution function to discrete distributions and failed.*

Power: *The comparison of the power of these tests is a complex issue. The power of a test depends greatly upon the nature of the alternative hypothesis. Three essential points can be prudently retained.*

1) The literature [DAT 10] seems to agree on the fact that in the absence of parameters to estimate, the AD^2 test is advantageous, as it presents more power.

2) Having parameters to estimate reduces the power difference between tests. The KS test, however rigorous, remains the least powerful compared to other tests.

3) The *CVM* family of tests shows more power distributions [DAT 10] than the KS test. This is hardly surprising since the KS test detects only large differences *at*

one point between the *empirical cumulative distribution* function and the *reference cumulative distribution* function. This constitutes a notable penalty. The members of the CMV family examine this difference across the entire domain of the variation of τ. In this respect, this test has an advantage over other tests [DAT 10].

2.5.9. *Shapiro–Wilk test of normality*

The Shapiro–Wilk [SHA 65] test, proposed in 1965, calculates a W-statistic which tests a random sample, $x_1, x_2, -, x_n$ coming specifically from a normal distribution. The W-statistical test is calculated below:

$$W_{sh} = \left(\sum_{i=1}^{n} a_i x_i\right)^2 \Big/ \left(\sum_{i=1}^{n} x_i - \bar{x}\right)^2 \qquad [2.54]$$

W_{sh} are the small values used in an initial normality test. Percentage points for the W statistic are obtained with the help of Monte Carlo simulations. The $x_i, x_2, ..., x_n$ are the ordered values of the sample (x_1 is the smallest). The test demands the regrouping of the first data. The real number of observations in each group is compared to the expected number of observations. The test statistic is calculated according to this difference. The number of groups and the way in which the belonging to these groups is defined will have an effect on the power of the test, i.e. the test is sensitive to the detection of starting points of the null hypothesis, H_0. Power is not only affected by the number of groups and the way in which they are defined, but also by the sample size and the form of the underlying null hypothesis of the distributions, which is *assumed to be true.* The continuous data present a pronounced difficulty and we define groups by segmenting the range of possible values without overlapping intervals.

Membership of a group can be defined by interval extremities. In general, power is maximized by the choice of parameters, in the same way as belonging to the group is equiprobable (i.e. the probabilities associated with an observation of falling into a given group are spread out as uniformly as possible within intervals). The rule suggests using the value (**$2n^{2/5}$**) as a starting point for choosing the number of groups. Given the pedagogical *importance* of the KS and χ^2 tests, in this chapter we will deliberately avoid using the Shapiro-Wilk test.

2.5.10. *Adequacy test of chi-square (χ^2)*

The χ^2 test is used to test [SNE 89] whether a sample of data coming from a population with a specific distribution equals the one used by the theoretical model.

This test does not work for small samples. An interesting characteristic of the test adjustment validity of χ^2 is that it can be applied to any *univaried* distribution for which we can calculate the cumulative distribution function. The Karl Pearson χ^2 of adjustment test validity is applied to data grouped by class (see Appendices, Volumes 2 and 3). However, the statistical value of the χ^2 test is dependent upon the way in which the data are put into *order*.

Another disadvantage of the χ^2 test is that a sample of sufficient size is required for the χ^2 approximation to be valid. This test is an alternative to the Anderson–Darling test and the KS test. The χ^2 test is appropriate for the adjustment of discrete distributions such as binominal distribution (Poisson). The following is defined for the H_0 and H_a hypotheses:

– ***H₀***: *Data which follow a specified distribution (null hypothesis);*

– ***Hₐ***: *Data which do not follow a specified distribution (alternative).*

To validate the (χ^2), the adjustment of the calculation data are divided into (k) classes and the text statistics are defined as:

$$\chi^2 = \sum_{i=1}^{k} \left(O_{(i)} - E_{(i)}\right)^2 \Big/ E_{(i)} \qquad [2.55]$$

where O_i is the observed frequency and where E_i represents the unexpected frequency. The test frequency is calculated using the following:

$$E_{(i)} = N\left\{F\left(Y_u\right) - F\left(Y_L\right)\right\} \qquad [2.56]$$

$F(.)$ is the cumulative distribution function of the tested distribution. $\boldsymbol{Y_u}$ is the superior limit of class $\boldsymbol{i}$. $\boldsymbol{Y_L}$ is the inferior limit for class $\boldsymbol{I}$. N is the sample size.

2.5.10.1. *Level of signification (α) or critical region*

The test statistic follows a χ^2 distribution with ($\kappa - s$) degrees of freedom, where κ is the number of parameters estimated (including the number of cells which are not empty) and c = the number of estimated parameters (including parameters of localization, of scale and of shape). For example, for a Weibull distribution with three parameters, $s = 4$. Consequently, the hypothesis that these data come from a population with a specified distribution is rejected if:

$$\chi^2 \succ \chi^2_{(\alpha,\kappa-1)} \qquad [2.57]$$

$\chi^2_{(\alpha,\kappa-1)}$ is the χ^2 function for 100 points with $(\kappa - s)$ degrees of freedom and a (α) level of signification. In the above formulae for the critical regions, the convention that follows χ^2_α is the superior critical value of the χ^2 distribution and $\chi^2_{(1-\alpha)}$ is the weakest critical value of the χ^2 distribution.

The critical values of χ^2 for different degrees of freedom are presented in the table in the appendices. This critical value is found at the point where a column (corresponding to a given probability) crosses a line (corresponding to degrees of freedom). For example, the critical value of χ^2 with $\nu = 4$ degrees of freedom for a probability of **0.25** is equal to **5.38527**. That means that the surface under the density curve of χ^2 with four degrees of freedom to the left of the value **5.38527** is equal to **0.25** (with 25% of the surface).

This test is applied in cases where one observes the cumulative distribution of (n) *components* in (κ) *classes*. It is used to test the hypothesis that the cumulative distribution of data is effectuated according to a theoretical distribution. It is a question of theoretical/practical adequacy. To test the adequacy of a theoretical cumulative distribution, we make use of two elements. On the one hand, (n) observations distributed in (κ) cells. That finishes with (n_i) being the number of observations in the i th cell. On the other hand, a theoretical distribution which fixes probability P_i $(i = 1, 2, 3, \ldots, \kappa)$ for each cell.

n_1	n_2	...	N_κ
P_1	P_2	...	$P\kappa$
nP_1	nP_2	...	$nP\kappa$

The *score* of a Pearson test constitutes a measurement of the distance between empirical cumulative distribution and theoretical distribution. It is based on the cumulative distribution of (n) *components* according to theoretical distribution. Subsequently, we calculate this test:

$$Test - Pearson_{observed} = \chi^2_{observed} = \sum_{i=1}^{\kappa} \frac{(n_i - np_i)^2}{np_i} \qquad [2.58]$$

The null hypothesis that we analyze using the Pearson test is H_0. The theoretical distribution is the true underlying distribution of the data. We can demonstrate the distribution of the χ^2 statistic using the H_0 hypothesis using a $\chi^2_{\kappa-1}$ distribution with $\nu = (\kappa - 1)$ degrees of freedom, if the expected number nPi is sufficiently high, i.e. ≥ 5. The null hypothesis H_0 is rejected if:

$\chi^2_{observed} \succ q\chi^2_{v-1}(at\ 95\%)$; $q\chi^2_{v-1}(95\%)$ *is the* quintile *at* 95% *of* χ^2_{v-1} *law*

Here, the principle of this test on the *distribution law of* χ^2 is illustrated:

$$For\ \tau = 0, 0.1, \ldots 10\ and\ v = 3 \rightarrow \chi^2(\tau, v) = \left(\frac{Exp\left\{-\frac{\tau}{2}\right\}}{2 \times \Gamma\left(\frac{v}{2}\right)} \right) \times \left(\frac{\tau}{2}\right)^{\left(\frac{v}{2}\right)-1}$$

In cases where the random variable is continuous, we must discretize areas where an element of ambiguity is introduced. In these cases, another test exists which uses empirical distribution and which is generally more powerful than the Pearson test. Although non-parametric tests, different from chi-square exist, they are presented in very different forms. They are all variants of a adequacy test relative to multi-nominal distribution.

χ2(τ, v) =
0.000
0.120
0.161
0.188
0.207
0.220
0.229
...

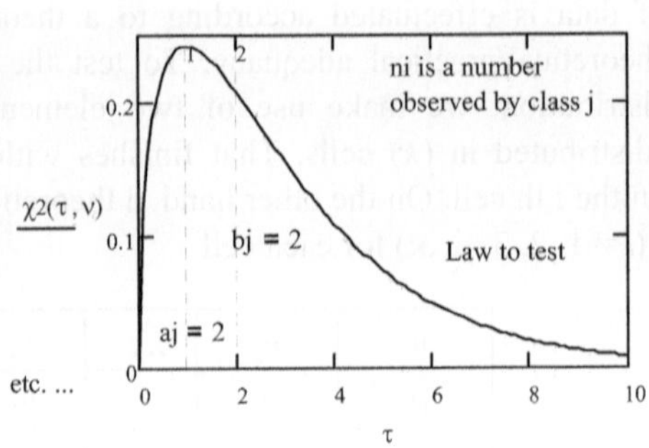

Figure 2.19. *Chi-square (χ^2) or Pearson test*

The *family* takes its name from the fact that the distribution of the χ^2 statistic under $\mathbf{H}_0$ (although generally unknown) always converges toward a χ^2 distribution for larger and larger samples ("asymptotic" distribution). The number of degrees of freedom of this limited distribution depends on the particular problem under consideration. χ^2 tests are always *asymptotic tests*. By Q, we indicate a statistic of χ^2:

$$Q = \sum_{i=1}^{\kappa} \frac{(n_i - np_i)^2}{np_i} \qquad [2.59]$$

Once the sample has been taken, the value of Q can be calculated. If by using O_i we indicate the observed size (*obs*), E_i the expected size (*exp*), the χ^2 statistic would be written:

$$Q = \sum_{i=1}^{K} \frac{(O_i - E_i)^2}{E_i} \qquad [2.60]$$

2.5.10.2. *Probability distribution function of the χ^2 distribution; dichisq (τ, ν)*

– **ν** is a positive integer called the degrees of freedom of the chi-square distribution.

– **τ** is a scalar or a vector (RV) of the chi-square distribution.

– **m** is an integer > 0 of the chi-square distribution.

Numerical application of χ^2 for correlation quality: *This* exercise allows us to execute a χ^2 test *(chi-squared)* to determine the quality of the correlation between the observed results and the "experimentally expected" results according to a probabilistic model. The *null* hypothesis indicates that the expected results correspond to the observations. To do this, we must enter the vectors of the observed frequencies and the expected frequencies:

exp =

	0
0	2.01
1	1.99
2	1.98
3	2.00
4	2.01
5	2.02
6	2.00
7	1.97
8	1.98
9	1.96

obs =

	0
0	1.98
1	2.00
2	1.98
3	2.02
4	2.00
5	2.03
6	1.98
7	1.97
8	2.00
9	1.96

χ2(τ, ν) =

0.00
$3.78 \cdot 10^{-10}$
$4.26 \cdot 10^{-9}$
$1.75 \cdot 10^{-8}$
$4.77 \cdot 10^{-8}$
$1.04 \cdot 10^{-7}$
$1.95 \cdot 10^{-7}$
$3.33 \cdot 10^{-7}$
$5.29 \cdot 10^{-7}$
$7.94 \cdot 10^{-7}$
...

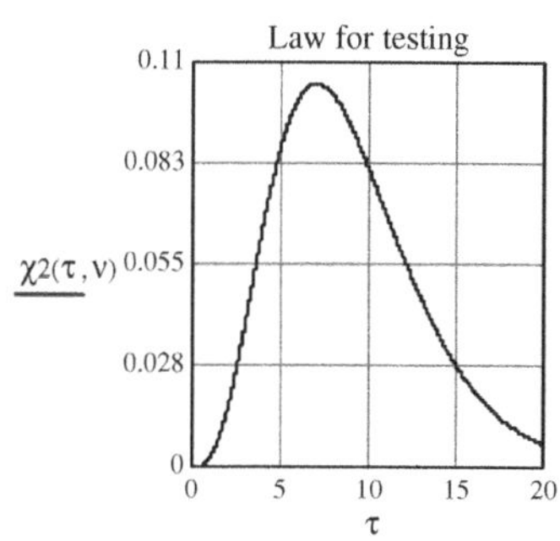

continuous ...

Figure 2.20. *Distribution tested by the Pearson (χ^2) test (chi-squared)*

The expected values of each category must be equal to at least five (5) for this analysis to be valid. Here, we are dealing with ten (10) values from a cylinder bore pierced to around 2 inches in diameter. The sum of the observed values must be equal to that of the expected values.

We calculate the respective sums: $\rightarrow \sum_{i=1}^{K} obs = 19.92 \; and \; \sum_{i=1}^{K} \exp = 19.92$

We enter the significance level (α) $\rightarrow \alpha = 5\%$

Degrees of freedom by the Chebyshev relation (ν) = $length(obs) - 1 = 10 - 1 = 9$

The χ^2 statistic allows: $\rightarrow \chi^2 = \sum \dfrac{\overline{(obs - \exp)^2}}{\exp} = 1.199 \times 10^{-3}$

$$P\left\{statistics\ of\ \chi^2 at\ least\ equal\ to\ P\right\} = P = 1 - pchisq\left\{\chi^2, \nu\right\} = 1$$

$$X_2\{Pt\ of\ percentage\ \alpha\} \rightarrow X^2 = pchisq\{1 - \alpha, \nu\} = 16.919$$

$$H_0\{\ null\ hypothesis\ rejected\ if\ \} \rightarrow X^2 \prec \chi^2$$

***Conclusion**: As $X^2 < \chi^2$, the null hypothesis (H_0) is rejected in this case.*

$\chi 2(\tau, \nu) =$

0
3.781×10^{-10}
4.256×10^{-9}
1.75×10^{-8}
4.767×10^{-8}
1.036×10^{-7}
1.951×10^{-7}
3.329×10^{-7}
...

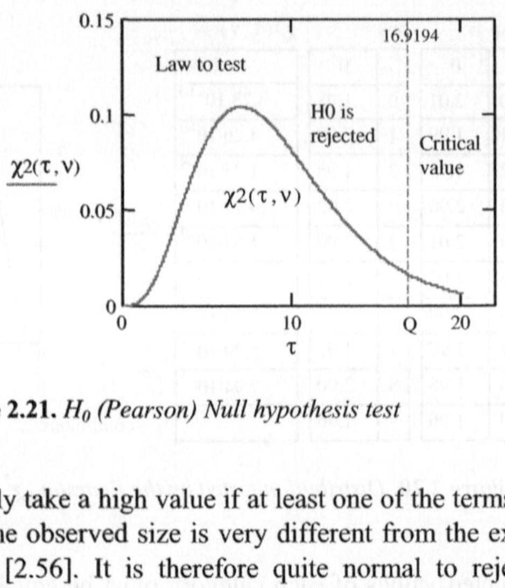

Figure 2.21. *H_0 (Pearson) Null hypothesis test*

DISCUSSION.– Q can only take a high value if at least one of the terms of the sum takes a high value, i.e. if the observed size is very different from the expected size for at least one modality [2.56]. It is therefore quite normal to reject the null hypothesis. Also, Q can only take a low value if all the terms of the sum have weak values, i.e. if all the sizes are close to their expected value. It is in this case, most definitely impossible to reject the null hypothesis. We can see that if all the n_i sizes are multiplied by m, the value of the Q statistic is equally multiplied by m. Thus, although the relationships between the sizes are unchanged, the null hypothesis becomes more difficult to accept. This reflects the fact that the quantity of information on a distribution rises with the size sample size. Many different variants on the χ^2 test exist.

2.6. Accelerated testing method

In fracture mechanics, the varying of a constraint is the equivalent to multiplying the failure rate by a constant. We say that we have true *physical acceleration*, sometimes called *true acceleration*. In mechanical reliability, that means that a unity of raised constraints (c.-to-d., raised temperature or tension, or humidity, or number of cycles etc.) of functioning shows the same failures that constraints of typical use would show, except that they would occur more rapidly, hence the term accelerated tests. Failure can be due to fatigue, corrosion, diffusion, migrations etc.

The same causes of failure exist under normal constraints. It is simply that the timescale is different. When there is true s*tress* acceleration, that is the equivalent of transforming the timescale used to record future failures. The transformations that are the most frequently used are linear. This means that the time for failure (due to raised stress) has just been multiplied by a constant that we call *factor of acceleration* (γ) to obtain the equivalent of the necessary time for failure resulting from the constraint of use. We will now present a factor of acceleration (γ) between *stress* and use allowing us to write:

Acceleration (γ) of linear relationships	
Time to reach failure	$t_u = (\gamma)\times t_s$
Probability of failure	$F_u(t) = F_s(t/(\gamma)$
Reliability	$R_u(t) = R_s(t/(\gamma)$
Distribution or density function	$f_u(t) = (1/(\gamma) f_s(t/(\gamma)$
Failure rate	$h_u(t) = h\ (1/(\gamma)_s(t/(\gamma)$

Table 2.8. *Factor of acceleration (γ)*

Where t_s is the time to fail due to stress

$F_s(t)$ is the cumulative distribution function of constraints (stress)

$f_s(t)$ is the distribution function of constraints

$h_s(t)$ is the rate of failure due to stress

Each model of failure has its own factor of acceleration (γ). That is made clear by the very fact that acceleration demands the existence of a physical process dependent on stress, thus provoking changes or degradations which would lead to failure (breakdown, damage). For example, for censored tests with time t_{je} the breakdown of the jth failure, we need an estimation of the cumulative distribution function. The simplest estimation is: $100 \times i/n$ (with a total of (n) units per test). That constitutes an overestimation, c.-to-d bias. Many texts recommend corrections like: $100(i-1/2)/n$ or $100 \times i/(n+1)$.

Let the reading time be T_1, T_2,..., T_k and let the failures correspond to the data recorded at each reading r_1, r_2, ..., r_k . There are (n) under testing (see Figure 2.21). On reading the time t_j, we use an estimation of the distribution function or the percentiles using the following expression:

$$\sum_{i=1}^{j} r_i \times 100 \Big/ n \qquad [2.61]$$

2.6.1. *Multi-censored tests*

The most complicated calculations are for the multi-censored data. The Kaplan-Meier (KM) estimation can be used to obtain the tracing positions of each time to fail. The modified KM estimation is recommended due to its precision. The ingenious idea is to take the distribution equation of the model and to write it in such a way that the function of $f(\tau)$ is a linear equation of (τ). For example, for an exponential model, we have:

$$\lambda \cdot \tau = \ln\left(\frac{1}{1-F(\tau)}\right) \; or \; \frac{\lambda \cdot \tau}{\ln(10)} = \log\left(\frac{1}{1-F(\tau)}\right) \qquad [2.62]$$

2.6.2. *Example of the exponential model*

If, for $y = 1/\{1- f(\tau)\}$ and $x = \tau$, then log (y) is linear on x with the gradient $(\lambda/\ln 10)$; that shows that we can trace the graph to logarithmic scale on the y axis. We use position tracing estimations for $F(\tau_i)$ described above (without the multiplier $\times 100$) to calculate the pairs of points (x_i, y_i) to trace. If the data are compatible with an exponential model, the line which results from it will have points on this line which line up almost in a straight line, passing by the origin with the gradient $(\lambda/\ln 10)$.

2.6.3. *Example of the Weibull model*

The Weibull distribution function model appears as:

$$\left\{ \begin{array}{l} \ln\ln\left(\dfrac{1}{1-F(\tau)}\right) = \gamma\ln(\tau) - \gamma\ln(\alpha) \; or \\ \\ \log\ln\left(\dfrac{1}{1-F(\tau)}\right) = \gamma\log(\tau) - \gamma\log(\alpha) \end{array} \right\} \qquad [2.63]$$

We postulate that $y = \ln [1/\{1-f(\tau)\}]$ and $x = \tau$. Log (y) are linear with the gradient (γ). We use habitual estimations of tracing position for $F(t_i)$ (without the multiplier 100) to calculate the pair (x_{je}, y_i) of points we want to trace. If the data are compatible with a Weibull model, the line which results will have the points of this straight line with the gradient (γ). This line will cross the data on the x axis with time $\tau = (\alpha)$ and the y axis with the intersection ($-\gamma$). It is, in fact, the Allan Plait diagram which is presented in the Appendix of Volume 2 for the Weibull distribution.

2.6.4. *Example for the log–normal model*

Let's postulate the distribution function for log–normal distribution as:

$$\left\{\ln(\tau) = \sigma.\Phi^{-1}\{F(\tau)\} + \ln T_{50} \text{ or } \log(\tau) = \frac{\sigma}{\ln 10}.\Phi^{-1}\{F(\tau)\} + \log T_{50}\right\} \quad [2.64]$$

with Φ^{-1} indicating the inverse function for normal distribution

Let us postulate $y = \tau$ and $x = \Phi^{-1}.\{f(\tau)\}$. ($y$) is linear with the gradient [$\sigma/\ln(10)$] and the intersection [when $f(\tau) = 1/2$] with T_{50}. We use habitual estimations of the tracing position for $F(\tau_i)$ (without the multiplier 100) to calculate the pairs of points (x_{je}, y_i) we want to trace. If the data are compatible with a log-normal model, the line which results from this will have the points on this line with a gradient of [$\sigma/\ln(10)$] which intercepts T_{50} on the y axis.

2.6.5. *Example of the extreme value distribution model (E-MIN)*

We consider the cumulative distribution function of the extreme value and we present:

$$\ln\{-\ln(1-F(\tau))\} = \left(\frac{x-\mu}{\beta}\right) \quad [2.65]$$

When $y = -\ln [1 - (\partial f/\partial \tau)]$, $\ln(y)$ is linear in x with a gradient of [$1/\beta$] which intercepts [$-\mu/\beta$], and we can use the paper to the semi-logarithmic scale on the axis. The points should follow a straight line with a gradient of [$1/\beta.\ln(10)$] and an interception of [$-\beta.\ln(10)$]. The factors of ln (10) are necessary.

2.6.6. *Example of the study on the Weibull distribution*

By using the data from the Weibull distribution, we have generated 20 times the failures with a shape parameter of $\gamma = 1.5$ and a MTBF = $\theta = 500$. Supposing the test time of T = 500 hours, only 10 components of these failures would have been observed. The failures take place at the following times: 54, 187, 216, 240, 244, 335, 361, 373, 375, and 386.

Here is a table which resumes these data:

Failure # i	Time of failure (x)	Estimation of F (t_i) (Je-.3)/20.4	ln {1/(1 − F (t_i) }(y)
1	54	.034	.035
2	187	.083	.087
3	216	.132	.142
4	240	.181	.200
5	244	.230	.262
6	335	.279	.328
7	361	.328	.398
8	373	.377	.474
9	375	.426	.556
10	386	.475	.645

Table 2.9. *Estimations of time to fails on a Weibull distribution*

In the Allan Plait paper (Appendix Figure A.2), we have considered the points of column (2) and of column (3). The graph is based on these data:

X	54	187	216	240	244	335	361	373	375	386
F	.035	.087	.142	.200	.262	.328	.398	.474	.556	.645

Table 2.10. *Data of (x) and (F) for the Weibull distribution trace*

The first points on a probability graph have more point variability in the central range. Less attention will be paid to them during the test. By linear regression, we have traced a line across the points on the Allan Plait paper. This has given us an estimation of the gradient to$1/\beta = 1{,}46$ close to the value 1.5 used in the simulation. The interception takes place at point (−4.114) which corresponds to the parameter $(-\gamma)$. We estimate then, that α = 657, the 'true' value. In the simulation α is equal to 500.

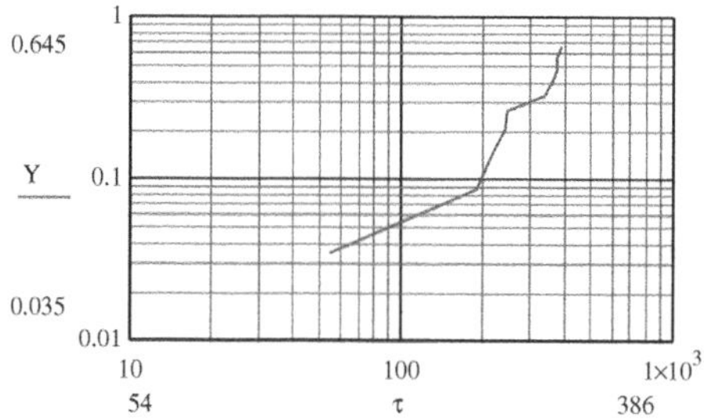

Figure 2.22. *Model of the extreme value distribution (E-MIN), Weibull*

If the data do not follow a normal distribution, we must use the COX model. Non-normality is a mode of lifespan with a whole part. This strategy aims to make the data of non-normal distributions close to data with normal distributions. Unfortunately, the choice of the best transformation is not generally obvious. In 1974, this problem was *solved* by what are known as BOX – COX [BOX 74, BOX 94]. These are defined uniquely for positive data values. That should not pose a problem, as a constant can always be added if the ensemble of observations contains one or several negative values. The box power transformations called BOX–COX are data by transformation, like those below:

$$x(\lambda)=\left(x^{\lambda}-1\right)/\lambda \ for\ \lambda\neq 0\ et\ x(\lambda)=\ln(\lambda)\ for\ \lambda=0 \quad [2.66]$$

Given the vector of observation data $\boldsymbol{x}$ = x_1 and x_2, ... x_n,, a way of selecting power (λ) is using the (λ) which maximizes the probability function logarithm (below):

$$\begin{Bmatrix} f(x,\lambda)=-\left(\frac{n}{2}\right)\ln\left(\sum_{i=1}^{n}\frac{\left[x_i(\lambda)-\bar{x}(\lambda)\right]^2}{n}\right)+(\lambda-1)\sum_{i=1}^{n}\ln(x_i) \\ \textit{With the arithmetic mean} \rightarrow \bar{x}(\lambda)=\frac{1}{n}\times\sum_{i=1}^{n}x_i(\lambda) \end{Bmatrix} \quad [2.67]$$

A confidence based on the statistic of the relationship of likelihood can be constructed on (λ) as shown below:

$$f(x,\lambda)\geq\left\{f\left(x,\hat{\lambda}\right)-\frac{1}{2}\chi^2_{\alpha,1}\right\} \quad [2.68]$$

Where $\hat{\lambda}$ indicates the *estimator of maximum likelihood* for (λ) and $\chi^2_{\alpha,1}$ is the superior percentile × 100 (1−α) of the χ^2 distribution with ν = 1 degree of freedom.

2.6.7. *Example of the BOX–COX model*

Data from the BOX-COX model						
.15	.09	.18	.10	.05	.12	.08
.05	.08	.10	.07	.02	.01	.10
.10	.10	.02	.10	.01	.40	.10
.05	.03	.05	.15	.10	.15	.09
.08	.18	.10	.20	.11	.30	.02
.20	.20	.30	.30	.40	.30	.05

Table 2.11. *Log-likelihood for the different values of λ. Source: [DAT 10]*

The logarithmic likelihood values (**LLV**), obtained by making (λ) vary from −2.0 to 2.0, are given below.

λ	LLV	λ	LLV	λ	LLV
−2.0	7.1146	−0.6	89.0587	0.7	103.0322
−1.9	14.1877	−0.5	92.7855	0.8	101.3254
−1.8	21.1356	−0.4	96.0974	0.9	99.3403
−1.7	27.9468	−0.3	98.9722	1.0	97.1030
−1.6	34.6082	−0.2	101.3923	1.1	94.6372
−1.5	41.1054	−0.1	103.3457	1.2	91.9643
−1.4	47.4229	0.0	104.8276	1.3	89.1034
−1.3	53.5432	0.1	105.8406	1.4	86.0714
1.2	59.4474	0.2	106.3947	1.5	82.8832
−1.1	***65.1147***	***0.3***	106.5069	1.6	79.5521
−0.9	75.6471	0.4	106.1994	1.7	76.0896
−0.8	80.4625	0.5	105.4985	1.8	72.5061
−0.7	84.9421	0.6	104.4330	1.9	68.8106

Table 2.12. *Logarithmic likelihood values function. Source [DAT 10]*

This table shows that $\lambda = 0.3$ maximizes the likelihood logarithm function (LLF). The Box–Cox method supplies a graphical method for the choice of (λ) to transform an ensemble of data into a normal distribution. The criteria used to choose (λ) for the linearity of the Box–Cox is the value of (λ) which maximizes the correlation between the transformed values of x and the y values at the time of the probability graph trace.

2.7. Trend tests

Statistical trend tests allow us to quantitatively determine whether the repair time of a system shows a significant trend, i.e. an improvement or a trend for damages. Let (r) be the system repairs produced with the system ages T_1, T_2, T_3, ... T_r (beginning of the observation period $T = 0$). Let $T = I_1$, $I_2 = T_2 - Tj_{i3} = T_3 - T_2$, ..., $Tj_{er} = T_r - T_{r-1}$ be the representation of the arrival of repair time (sequence of response time between two breakdowns). Let us suppose that the observation period ends at the moment $T_{final} > T_r$. We will calculate the number of occurrences of what are known as the *inversions* of failure. If there are many repetitions, this significantly proves an improvement trend, otherwise it is proof of damage. We proceed in the following way:

– We count a reversal every time that $je_j < I_k$ with $j < k$.

– The inversion count is the total number of repetitions $\boldsymbol{R}$.

– For repair time r, the maximum possible number of repetitions is worth $(r-1)/2$.

– If there are no mean trends, we predict: $r\,(r-1)/4$ repetitions.

For example, we have five system repairs at ages 22, 58, 71, 156 and 225 (h). The observation period came to an end in the time of 300 (h). First of all, let us calculate the time of the arrival of maturity to obtain: 22, 36, 13, 85, 69 (h). Next, we calculate the turnaround number by indicating the first component to arrive at failure in 22 hours, and count the number of late arrivals. There were three late arrivals. In this way, for the second time after 36 hours, there were exactly two. We have obtained two with zero inversions. The total will then be $R = 3 + 2 + 2 + 0 = 7$. The total number of possible reversals is $5 \times 4/2 = 10$ whose *mean* is five. In this example, we have had seven repetitions (two more than the mean). The table below shows that at the confidence interval of 90%, 95%, or 99%, we obtain the proof of the improvement of trend.

Value of R indicative of the significant improvement for the unilateral test			
Number of repairs	**R** Minimum for **90%** proof of improvement	**R** Minimum for p **95%** proof of improvement	**R** Minimum **99%** proof of improvement
4	6	6	–
5	9	9	10
6	12	13	14
7	16	17	19
8	20	22	24
9	25	27	30
10	31	33	36
11	37	39	43
12	43	46	50

Table 2.13. *Reliability test (Unilateral test). Source: [DAT 10]*

2.7.1. *A unilateral test*

A unilateral test is made up of a few means in cases where we know the measures taken to improve reliability. For $r = 5$ repairs (see example above) to occur where we have $R = 7$, the table (2.14) shows that we do not (yet) have enough proof to show significant improvement. That does not signify that an improvement model is incorrect, it simply signifies that there is not yet statistical "proof". With a small number of repairs, it is not easy to obtain significant results. For more than 12 repairs in a collection of data, we use this formula:

$$Z(R,r) = \left(R - \left(\frac{r(r-1)}{4}\right) + \frac{1}{2}\right) \Big/ \sqrt{\frac{(2r+5) \times r(r-1)}{72}} \qquad [2.69]$$

Z is a standard normal distribution. If $Z > 1.282$, we have an importance to at least 90%. If $Z > 1.645$, we have it to 95% and a $Z > 2.33$ indicates a significance to 99%. If, on the other hand, there is a trend leaning toward damage (fatigued or overloaded system), to determine whether the data confirm the trend, we expect a weak value for R. Table R indicates whether the value is sufficiently weak to be significant.

Unilateral test, values of *R* degradation indicating significance			
Number of repairs	*Maximum R for proof of 90% degradation*	*Maximum R for 95% proof of degradation*	*Maximum R 99% proof of degradation*
4	0	0	–
5	1	1	0
6	3	2	1
7	5	4	2
8	8	6	4
9	11	9	6
10	14	12	9
11	18	16	12
12	23	20	16

Table 2.14. *Unilateral test, degradation indicating trend [DAT 10]*

For the number of repairs $r > 12$, we use the above approximation, with R replaced by $\{r\ (r-1)/(2-R)\}$. This trend test is recommended due to the success of the model whose simple illustration for $R = 7$ and $t\ r$ (2–20) is presented here.

Another test "*Military Handbook*" allows us to find a significance when choice operates between absence and trend and the Power law (Duane model).

Z(R,r) =

14
6.267
3.057
1.225
0
–0.901
–1.608
–2.189
–2.683
–3.114
–3.497
–3.844
...

$$Z(R,r) = \frac{\left[R - \left[\frac{r\cdot(r-1)}{4}\right] + \frac{1}{2}\right]}{\sqrt{\frac{(2\cdot r+5)\cdot r\cdot(r-1)}{72}}}$$

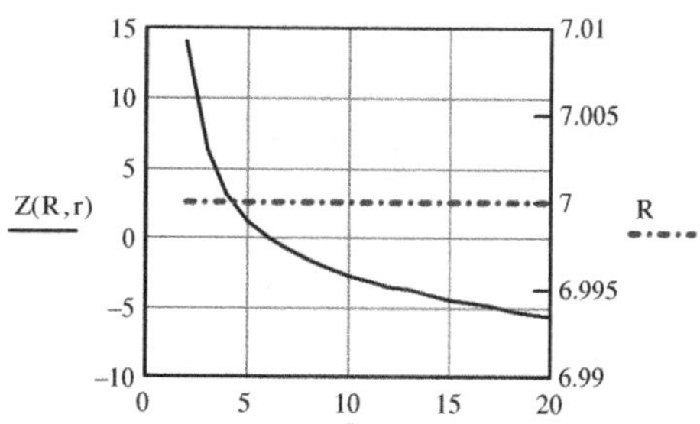

Continuous

Figure 2.23. *Unilateral test for trend and reliability*

2.7.2. *The military handbook test (from the US Army)*

The "Military Handbook" test comes from the United States. If the data comes from a system which follows the Power Law (e.g. exponential), this test will be significant. Let us take (r) as an example for repair times T_1, T_2, T_3, ..., T_r on the observation period which leans toward $T_{final} > T_r$. We calculate the following using the Pearson criterion χ^2 with $2r$ degrees of freedom:

$$\chi_{2r}^2 = 2 \times \sum_{i=1}^{r} \ln\left(\frac{Trend}{T_i}\right) \qquad [2.70]$$

Next, we will compare the result to the percentiles of the χ^2 distribution with $2r$ degrees of freedom. For a trial of unilateral improvement, we do not reject any trends in favor of an improvement trend if the value of χ^2 is superior to 90 (even 99) percentiles. For a test of unilateral degradation, we do not reject any trend if the value of χ^2 is inferior to a centile of 10 or less.

2.7.3. *The Laplace test*

This test is preferable to significance when the choice is presented between absence and trend and a function from an exponential mode. Using as an example repairs (r) with the times: T_1, T_2, T_3, ... T_r on the observation period ending at instant $T_{end} > T_r$. We calculate $Z(r)$:

$$Z(r) = \sqrt{12r} \times \sum_{i=1}^{r}\left(T_i - \frac{Trend}{2}\right) \Big/ r \times Trend \qquad [2.71]$$

Let us now compare the result with the largest centile (improvement) or with the weak centile (degradation) from the standard normal distribution centiles. Let us take the percentile = 100× $f(Z)$, which is the probability distribution function.

2.7.4. *Homogenous Poisson Process (HPP)*

The simplest model is represented by $M(\tau) = (\lambda.\tau)$. For a constant repair rate, we postulate $m(\tau) = (\lambda)$. This model is delivered at the moment when the time between two independent and identically distributed breakdowns is distributed according to an exponential law of parameter (λ). This basic model is equally known as a homogenous Poisson process. The following classic formulae apply:

$$F(\tau) = 1 - Exp\{-\lambda.\tau\} \qquad [2.72]$$

This is the cumulative distribution function of the expectancy for the next failure to occur. $N(T)$ is the cumulated number of failures from time $T_1 = 0$ to time T. We therefore present:

$$P\{N(\tau)=k\}=\frac{(\lambda\tau)^k \times Exp\{-\lambda.\tau\}}{k!} \qquad [2.73]$$

– $M(T) = (\lambda.T)$, failure rate expected in time (T)

– $M(\tau) = m(\tau) = \lambda$ which is the repair rate

– $(1/\lambda)$ is the MTBF (mean time before failure)

Usually, the flat repair portion corresponds to the type of functioning whose lifespan is relatively constant. In mechanical reliability, the Poissonian process is the model which best represents this portion of the curve. The HPP model deals with the flat part of the bathtub curve. It is largely used for the equipment of reparable systems. That comes, in part from the form of the empirical bathtub curve below:

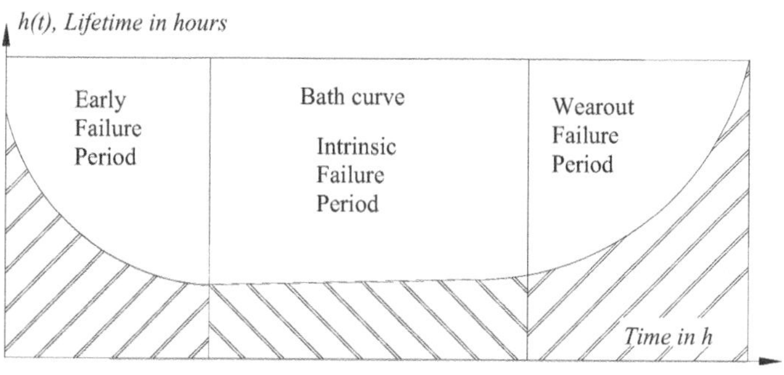

Figure 2.24. *Bathtub curve in classical probability*

We can see that in the HPP model, the probability of having exactly k failures with time T is given by the Poisson distribution with mean T. With the help of a simple Excel table, we postulate:

Y = POIPDF (k, λ T) or by the expression EXCEL: POISSON (k, λt , FALSE)

2.7.4.1. *Rate of repairs of the Non-Homogenous Poisson Process (NHPP)*

A flexible model with notable success in many applications, for the rate of failure on time τ to begin with, $M(\tau)$, is given by the polynomial:

$$M(\tau)=a\times\tau^{b} \;\; for \;\; (a,b)\succ 0 \qquad [2.74]$$

We notice, on looking at Figure 2.25(a) that the rate of failure is increasing:

$$m(\tau) = a \cdot b.\tau^{b-1} = \alpha.\tau^{-\beta} \ for\ \alpha \succ 0\ and\ \beta \prec 1 \qquad [2.75]$$

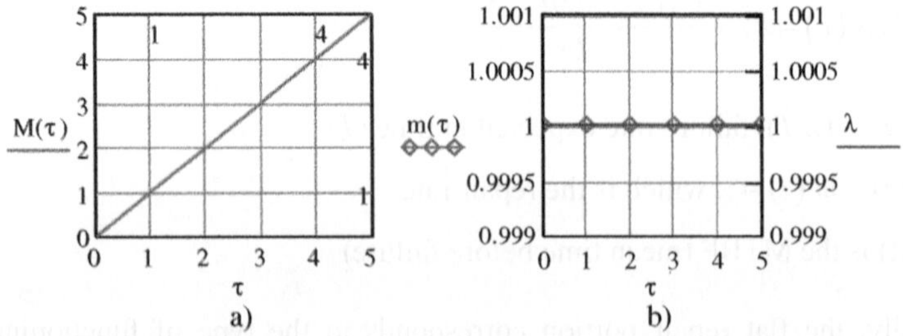

Figure 2.25. *a) Repair rate for the Non-homogenous Poisson Process (NHPP). b) Repair rate for the NHPP. λ = Constant*

The Power Law model is flexible and considers the *HPP* as a special case. The Poisson process has a constant repair rate model $m\ (\tau) = \lambda$ (Figure 2.25(b)). Substituting the function $\lambda\ (\tau)$ with (λ), we will have a *NHPP* with λ and with an intensity function $\lambda(\tau) = m(\tau) = \alpha.\tau^{-\beta}$. Due to the polynomial nature of the repair rate, this model is flexible and allows the simultaneous modeling of growth ($b > 1$ or $\beta < 0$). While decreasing ($0 < b < 1$ or $0 < \beta < 1$) it expresses the failure rate. When $b = 1$ or $\beta = 0$, the model is reduced to the HPP model of repairs with a constant speed. The failure probabilities of all the NHPPs can be calculated using the Poisson formula.

$$P\{N(T) = k\} = \frac{M(T)^k}{k!} E\,\mathrm{xp}^{-M(\tau)} \qquad [2.76]$$

The probabilities of a number of failures for the NHPP model are calculated by a pure and simple calculation of the formula for the HPP as well as for all NHPPs. For the Power Law model, we postulate:

$$P\{N(T) = k\} = \frac{\left(aT^b\right)^k}{k!} \exp^{-\left(aT^b\right)} \qquad [2.77]$$

The Power law model is also called the *Duane Model*. This model as well as the **AMSAA** model (*Army Material Systems Analysis Activity*, 1970). The time-to-fail for a distribution process of the Power law has a Weibull distribution with β a shape parameter. The title of *Weibull Process* brings confusion, as it confuses lifespan distribution models applicable to the lifespan of a non-reparable population and a model of the time that it takes for a population to arrive at the failure of a reparable population. For all NHPPs, the distribution function $f(\tau)$ with the arrival time (τ) for the following failure, from the fact that the next failure occurs at instant (T) is given below:

$$F_T(\tau) = 1 - Exp\left\{-\int_0^{\tau} m(T+\tau)\,d\tau\right\} \quad [2.78]$$

When a breakdown occurs, the response time for the next failure, for a NHPP, is expressed by the cumulative distribution function $F_T(\tau)$. For Power Law response time with failure time T:

$$F_T(\tau) = 1 - Exp\left\{-a\left((T+\tau)^b - T^b\right)\right\} \quad [2.79]$$

If the Power Law applies, the rate of repair improvement, in time, is written: $[\alpha.\tau^{-\beta}]$. The exponent (β) or the gradient of growth of reliability will be included between [0 and 1]. The number of failures in time (τ) takes the form $\tau = M^b$ and the rate of repair takes the form (τ) $= m^{b-1}$. This will be a model of improvement when $0 < b < 1$. It would be practical to define $\beta = (1-b)$ and $\alpha = (ab)$ and to write the expression of the repair rate in this way: $m(\tau) = \alpha\tau^{-\beta}$.

We have improvement when $0 < \beta < 1$ with (β) which is known as being in the Duane domain [DUA 64]. It expresses the gradient or the improvement of reliability *(growth gradient)* in terms of original parameters for $M(\tau)$, we have:

$$a = \left(\frac{\alpha}{1-\beta}\right) \text{ and } b = (1-\beta) \quad [2.80]$$

This model of Power law from the growth postulates the following:

– *During the test in progress, system improvements are introduced. This produces continuous improvement on system repair rates.*

– *With a sufficiently long period of time, the effect of these improvements is modeled by the polynomial model of repair of continuous improvement rates* $[\alpha.\tau^{\beta}]$.

– *At the end of improvement, the MTBF is constant with its last expected value.*

When the improvement test ends at instant T and there are no improvements in progress, the repair rate remains constant $[\alpha \times T^{-\beta}]$. The MTBF of the system is inverse to the final repair rate $[(T^{\beta})/\alpha]$. If we estimate that the number of failures predicted until instant T by the real number observed, the estimated MTBF at the end of the reliability test of power law is written:

$$MTBF_{estimated\_end\_}test = \left(\frac{T}{r(1-\beta)}\right) \quad [2.81]$$

– *T is the duration of the trial.*

– *(r) is the total number of trial failures.*

– *(β) is the reliability growth slope.*

2.7.4.2. *Simulated example of the NHPP data*

1) Enter the constant positive users *a* and *b*.

2) Simulate a uniform vector of $n \in [0,1]$ of random numbers and call them U_1, U_2, U_3, . . . U_n.

3) Calculate $Y_1 = \{-1/a^* \ln U_1\}$ ** $1/b$.

4) Calculate $Y_i = \{(Y_{i-1}$ ** $b)$ $-1/$ a * $\ln U_i\}$ ** $1/b$ for $i = 2, \ldots, N$.

The *n* numbers $Y_1, Y_2, \ldots, Y_n$ are the desired, simulated times of repair based on the NHPP with the parameters *a, b* (or $\beta = 1-b$ and $\alpha = a \times b$). Let the function be *powersim.dp* which is used to generate 13 times the random repair of the NHPP process of Power Law with $U = 0.2$ and $b = 0.4$.

– *Let us enter the number N of simulated desired repair times* = *13*;

– *Let us enter the value of shape parameter (U > 0)* = *0.2*;

– *Let us enter the shape parameter b (b > 0)* = *0.4*.

2.8. Duane model power law

The Duane model distribution is HPP with an expected failure number $M(\tau)$, and a repair rate, $M'(\tau)$ = $m(\tau)$, given by respective expressions from [2.71]. The parameter $\beta = (1-b)$ is called the reliability growth gradient: 0.3 to 0.6. If a system is observed on a fixed time T and fails arise at moments $\tau_1, \tau_2, \tau_3, \ldots, \tau_r$ (beginning of the trial period or observation period with time 0), a Duane *parcel* is a (τ_i/i) parcel compared with t_i to the log–log scale (Figure 2.26). If the data follow a Power law model, the points are situated on the straight line with a gradient β and an interception ($\tau = 1$).

FractureNum =

	0
0	1
1	2
2	3
3	4
4	5
5	6
6	7
7	8
8	9
9	10
10	11
11	12
12	13

FractureTime =

	0
0	26
1	182
2	321
3	728
4	896
5	$1.268 \cdot 10^3$
6	$1.507 \cdot 10^3$
7	$2.325 \cdot 10^3$
8	$3.427 \cdot 10^3$
9	$1.187 \cdot 10^4$
10	$1.198 \cdot 10^4$
11	$1.356 \cdot 10^4$
12	$1.505 \cdot 10^4$

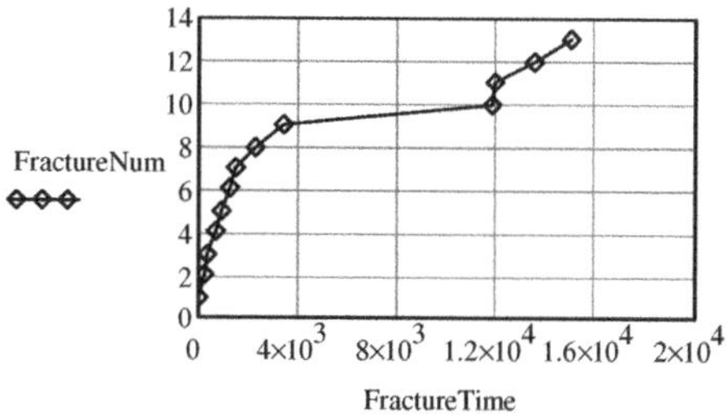

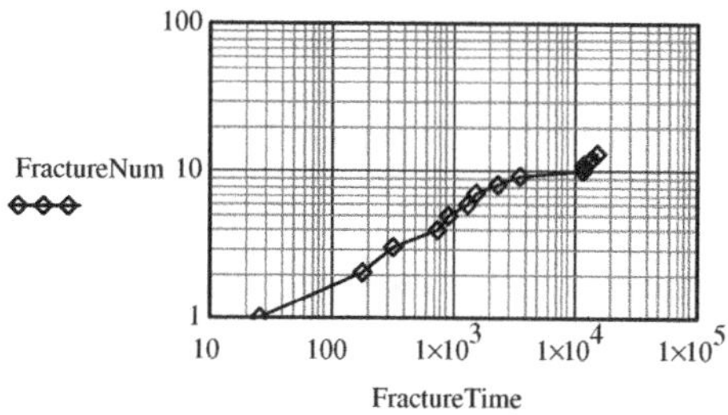

Figure 2.26. *Time-to-fail for a NHPP*

Graphical estimations come from a regression adjustment of $Y = \ln(\tau_i/i)$, according to $X = \ln \tau_i$. The gradient corresponds to (β). The estimation of b is $(1-\beta)$. These estimations are also modified by the *maximum of likelihood*. The formulae are given for a fixed T duration and (r) failures arising at times $\tau_1, \tau_2, \tau_3, \ldots, t_r$.

$$\hat{\beta} = 1 - \frac{r-1}{\sum_{i=1}^{r} \ln\left(\frac{T}{\tau_i}\right)}; \ \hat{a} = \left(\frac{r}{T^{1-\hat{\beta}}}\right); \ \hat{b} = 1 - \hat{\beta} = \frac{r-1}{\sum_{i=1}^{r} \ln\left(\frac{T}{\tau_i}\right)} \quad [2.82]$$

The MTBF estimated at the end of the test (or observation) period is:

$$MTBF_{estimated\_end\_test} = \left(\frac{T}{r \times \left(1-\hat{\beta}\right)}\right) = \left(\frac{T}{r \times \hat{b}}\right) \quad [2.83]$$

Approximate confidence limits for the MTBF at the end of the test: we postulate $100 \times (1-\alpha)\%$ confidence interval (M_L, M_U) for the MTBF at the end of the test. We can see that M_L is written $100 \times (\alpha/2)$ and M_U $100 \times (\alpha/2)$. We can:

$$\left\{\begin{array}{l} M_L = MTBF \times \dfrac{r-1}{\left(r + \dfrac{Z^2_{\alpha/2}}{4} + \sqrt{r \times \dfrac{Z^2_{\alpha/2}}{2} + \dfrac{Z^4_{\alpha/2}}{16}}\right)^2} \\ \\ and \;\; M_U = MTBF \times \dfrac{r(r-1)}{\left(r - Z_{\alpha/2} \times \sqrt{\dfrac{r}{2}}\right)^2} \end{array}\right\} \quad [2.84]$$

with $Z_{\alpha/2} > 100 \times (1 - \alpha/2)$ percentile points of standard normal distribution.

2.9. Chi-Square test for the correlation quantity

This exercise allows the execution of a chi-square test to determine the quality of the correlation between the observed results and the "experimentally expected results" according to a probabilistic model. The null hypothesis is that the expected results will correspond to observations. To do this, we enter the observed and expected vectors and frequencies and follow the seven steps below:

$$Obs = \{\ 1.98 \quad 2.0 \quad 1.98 \quad 2.02 \quad 2.0 \quad 2.03 \quad 1.98 \quad 1.97 \quad 2.0 \quad 1.96\ \}$$

$$Exp = \{\ 2.01 \quad 1.98 \quad 1.99 \quad 2.0 \quad 2.01 \quad 2.02 \quad 2. \quad 1.97 \quad 1.98 \quad 1.96\ \}$$

These expected values in each category must be at least equal to five for this analysis to be valid. Here, we will treat 10 values of around 2 inches. The sum of the observed values must be equal to that of the expected values.

1) Expected sums of Σ obs and Σ exp: $\left\{\sum obs = 19.92 \text{ and } \sum \exp = 19.92\right\}$

2) Enter the significance level = $\alpha = 0.05$;

3) Degrees of freedom v: $\nu = \{lenght(obs) - 1\} = (10 - 1) = 9$;

4) Chi-square statistic χ^2: $\chi^2 = \sum \overline{\dfrac{(obs - exp)^2}{exp}} = 1.119 \times 10^{-3}$;

5) P (statistic χ^2 at least equal to): $P = \left\{1 - pchi - square\left(\chi^2, \nu\right)\right\} = 1$;

6) Percentage point α: $X^2 = \{qchi - square(1-\alpha, v)\} = 16.919$;

7) If $X^2 < \chi^2 = 0$, null hypothesis is accepted as $16.919 > 1.119 \times 10^{-3}$ *(QED).*

2.9.1. *Estimations and χ^2 test to determine the confidence interval*

This exercise allows the execution of a *Chi-square* test to determine the confidence interval. To do this, we enter the data into the respective confidence thresholds:

– number of parts observed, $n = 10$;

– truncation of the number of failures, $\kappa = 4$ and 3;

– censoring of the number of failures, $r = 4$;

– duration of the experiments (trials) after which the trial is either censored or truncated, $T = 300$ hours (in our case study);

– duration of total functioning of the mechanism (ensemble of components), T_f that we calculate only in the case of non-replacement of failing components. If the test is realized with a replacement, it is suitable to refer to the theory developed in Chapter 1, i.e. we calculate $T_f = (n \times T)$;

– τ_i is the trial duration until the i th failure.

Let us take: $n = 10,\ \kappa = 3,\ r = 4,\ \tau_i = i = 100, 200, \ldots\ until\ 1200,\ T = 300h$

2.9.1.1. *Truncated tests*

$$T_f(i) = \left\{ (n-\kappa) \times T + \sum_{i=1}^{\kappa} \tau_i \right\} = 2.1 \times 10^3 \text{ hours}$$

For $\kappa = 0$, we will calculate the estimators corresponding to a probability of *a* ½:

$$\hat{\theta} = \left(\frac{T_f(i)}{0.69} \right) = \left(\frac{2.1 \times 10^3}{0.69} \right) = 3.043 \times 10^3 \text{ hours}$$

Otherwise $\kappa = 3$: $\hat{\theta} = \left(\dfrac{T_f(i)}{3} \right) = \left(\dfrac{2.1 \times 10^3}{3} \right) = 1.014 \times 10^3$ hours

2.9.1.2. *Censored tests*

$$T_f(i) = \left\{(n+1-r)\times\tau_i + \sum_{i=1}^{r-1}\tau_i\right\} \text{ for } \kappa = 4,\ \hat{\theta}(i) = \frac{T_f(i)}{4}$$

$$T_f(i) = \begin{pmatrix} 0.7\cdot10^3 \\ 2.1\cdot10^3 \\ 3.5\cdot10^3 \\ 4.9\cdot10^3 \\ 6.3\cdot10^3 \\ 7.7\cdot10^3 \end{pmatrix},\ \hat{\theta}_{\substack{censored \\ qd\ \kappa=0}} = \begin{pmatrix} 1.014\cdot10^3 \\ 3.043\cdot10^3 \\ 5.072\cdot10^3 \\ 7.101\cdot10^3 \\ 9.130\cdot10^3 \\ 1.116\cdot10^4 \end{pmatrix},\ \frac{\hat{\theta}_{censored}}{4} = \begin{pmatrix} 253.6230 \\ 760\cdot8700 \\ 1.268\cdot10^3 \\ 1.775\cdot10^3 \\ 2.283\cdot10^3 \\ 2.790\cdot10^3 \end{pmatrix}$$

– χ^2 theory for a *truncated test* with 2*v* with α possibilities of being exceeded

$$\left\{\frac{2\times Tf}{\chi^2_{\{2(\kappa+1),\ \alpha_2\}}}\right\} \prec \theta \le \left\{\frac{2\times Tf}{\chi^2_{\{2\kappa,\ 1-\alpha_2\}}}\right\}$$

– χ^2 theory for a *censored test* with $\{1-(\alpha_1+\alpha_1)\}$:

$$\left\{\frac{2\times Tf}{\chi^2_{\{2r,\ \alpha_2\}}}\right\} \prec \theta \le \left\{\frac{2\times Tf}{\chi^2_{\{2r,\ 1-\alpha_1\}}}\right\}$$

2.9.1.3. *Truncated tests at the confidence level*

With $\{1-(\alpha_1+\alpha_2)\}$ *and for* $v_{trunc-Obs} = 5$, *with* $v_2 = 9$

Degrees of freedom v_1: $v_1 = 2(\kappa+1) = \{v_{Trunc-Obs} - 1\} = 4$.

Knowing that $\alpha_1+\alpha_2 = 1$, we postulate:

At the confidence threshold $\alpha_2 = 0.025$, *with* $\alpha_1 = 1-\alpha_2 = 0.975$

$\chi^2 = qchisquare(1-\alpha_2, v_2) = 19.023$ *and* $\chi^2 = qchisquare\ x(1-\alpha_1, v_1) = 0.484$

2.9.1.4. *Chi-Square theory, truncated tests with* $2v$ *with* α *possibility of being exceeded*

Truncation→ $T_f = 1500^{th}\ hour \rightarrow \left\{\dfrac{2\times Tf}{\chi^2_{\{2(\kappa+1),\ \alpha_2\}}}\right\} \prec \theta \le \left\{\dfrac{2\times Tf}{\chi^2_{\{2\kappa,\ 1-\alpha_2\}}}\right\}$ we will have

$$\left\{\frac{2\times Tf(i)}{q\ chis-quare(1-\alpha_2,\nu_2)}\right\} \prec \theta \le \left\{\frac{2\times Tf(i)}{q\ chi-square(1-\alpha_1,\nu_1)}\right\} \Rightarrow 157.706h \prec \theta \le 6.193\times 10^3\ h$$

2.9.1.5. *Case of censored tests at confidence level*

With $\{1-(\alpha_1+\alpha_2)\}$ *and for* $\nu_{CensObs} = 10,$ *with* $\kappa = 4$

Degrees of freedom (*df*) $\nu_1 :\ \nu_1 = 2(\kappa+1) = \{\nu_{CensObs} - 1\} = 9$ (see Appendices, Table 9, Volumes 2 and 3).

At the confidence threshold $\alpha_2 = 0.025,$ *with* $\alpha_1 = 1-\alpha_2 = 0.975$ (see Appendices, Table 9, Volumes 2 and 3).

$$\chi^2 = q\ chi\ square(1-\alpha_2,\nu_2) = 19.023\ and\ \chi^2 = q\ chi\ square(1-\alpha_1,\nu_1) = 2.7$$

2.9.1.6. *Chi-square* χ^2 *theory, censored tests with 1−* $[\alpha_1 + \alpha_2]$

Censored test to $T_f = 2500^{th}\ hour \rightarrow \left\{\dfrac{2\times Tf}{\chi^2_{\{2r,\ \alpha_2\}}}\right\} \prec \theta \le \left\{\dfrac{2\times Tf}{\chi^2_{\{2r,\ 1-\alpha_1\}}}\right\}$, we have:

$$\left\{\frac{2\times Tf(i)}{q\ chi\ square(1-\alpha_2,\nu_2)}\right\} \prec \theta \le \left\{\frac{2\times Tf(i)}{q\ chi\ square(1-\alpha_1,\nu_1)}\right\} \Rightarrow 262.843h \prec \theta \le 1.852\times 10^3\ h$$

We have just calculated the χ^2 respective to truncated and censored tests. Next, we have framed confidence intervals, thus justifying the limits of use of components with pre-established intervals in the hypothesis $\chi^2_{2r,\alpha}$ with (2*r*) *df* having (α) to be exceeded. The literature sometimes proposes graphical solutions with a functional scale. In our opinion, these approaches are marred by uncertainties which are highly unjustified. For this reason, we give priority to the calculations.

2.9.2. *t_test of normal mean*

What is shown below is an example of a *t_test* relative to the hypothesis on two normal populations of the same mean. We are going to use the program MathCAD to solve the problem. To do this, we will enter two vectors data to analyze:

Ø1 =

	0
0	1.995
1	1.995
2	1.990
3	2.001
4	2.005
5	2.000
6	2.000
7	1.999
8	2.001
9	2.000
10	2.015
11	2.002
12	2.003
13	2.001
14	2.000

Ø2 =

	0
0	2.000
1	2.005
2	2.001
3	2.000
4	1.955
5	2.000
6	2.015
7	2.004
8	2.000
9	1.995
10	2.000
11	2.015
12	2.000
13	2.005
14	2.000

Problem presented :

Reject or accept the null hypothesis H_0?

We enter (a), a threshold of significance.

Let: ($\alpha = 0.01$)

Sample size, n_1 and n_2

– n_1= *length (Ø1) corresponding to* $n_1 = 15$ *values of Ø measured in inches*

– n_2= *length (Ø2) corresponding to* $n_2 = 15$ *values of Ø measured in inches*

Means, μ

– $m_1 = \mu_1$ = *mean (Ø1) corresponding to* m_1 = *mean of Ø measured* = 2

–$m_2 = \mu_2$ = *mean (Ø2) corresponding to* m_2 = *mean of Ø measured* = 2

Standard deviations $\sigma = s$

$$s_1 = stdev(\varnothing 1) \cdot \sqrt{\frac{n_1}{n_1 - 1}} = 5.436 \times 10^{-3} \text{ and } s_2 = stdev(\varnothing 2) \cdot \sqrt{\frac{n_2}{n_2 - 1}} = 0.014$$

Degrees of freedom v: So $v = n_1 + n_2 - 2 = 15 + 15 - 2 = 28$

2.9.3. *Standard error of the estimated difference, s*

$$s_1 = stdev(\varnothing 1) \cdot \sqrt{\frac{n_1}{n_1 - 1}} = 5.436 \times 10^{-3} \text{ and } s_2 = stdev(\varnothing 2) \cdot \sqrt{\frac{n_2}{n_2 - 1}} = 0.014$$

Degrees of freedom v: So v $= n_1 + n_2 - 2 = 15 + 15 - 2 = 28$

Standard error of the estimated difference, s:

$$s = \sqrt{\frac{(n_1 - 1) \times s_1^2 + (n_2 - 1) \times s_2^2}{v} \left(\frac{1}{n_1} + \frac{1}{n_1} \right)} = 3.766 \times 10^{-3}$$

– **Statistical test, *t***

$$t_1 = \left(\frac{m_1 - m_2}{s} \right) = 0.212 \text{ and } \mathbf{P[|T|>|t|]}\text{: Then } P = 2 \times \left\{ 1 - pt\left(|t_1|, v \right) \right\} = 0.833$$

***α*/2 Percentage point**: $T = \left| qt\left(\frac{\alpha}{2}, v \right) \right| = 2.763$

– **Rejection of the null hypothesis if |*t*| > *T***: So $|t_1| \succ T \rightarrow and\ T = 0$

2.10. Chebyshev's inequality

Sometimes, the probabilistic model is totally unknown. However, we know that to model a physical phenomenon, it is imperative that we know the closest probabilistic model possible. By probabilistic model, we mean a model characterized by its mean and its variance (or its standard deviation). Chebyshev showed, by his audacious idea that we can obtain, by a probability limit that the RV (random variable) is situated in an interval centered on the mean. According to what is presented above, we postulate:

$$P\{(\mu-\tau.\sigma)\leq X\leq(\mu+\tau.\sigma)\}\geq\left(\frac{\tau^2-1}{\tau^2}\right),\ \tau\succ 0 \qquad [2.85]$$

Thus, it is possible to see that the RV(X) is made up in the interval $X\subset\{(\mu-\tau.\sigma)$ et $(\mu+\tau.\sigma)\}$ is at least equal to $\left(\tau^2-1\right)/\tau^2$.

$$\text{Let: } P\{X\prec(\mu-\tau.\sigma)\ and\ X\succ(\mu+\tau.\sigma)\}\leq\left(\frac{1}{\tau^2}\right),\ \tau\succ 0 \qquad [2.86]$$

With τ a positive value, $\tau > 0$. For example, for $\tau = 3$, we will have:

$$P\{\mu-\tau.\sigma\leq X\leq\mu+\tau.\sigma\}\geq\left(\frac{\tau^2-1}{\tau^2}\right)\succ\left(\frac{9-1}{9}\right)\succ 0.889$$

We devised a simple workshop which allows us to calculate this probability for τ variant between 1 and 20. The results and the graphical interpretation of Chebyshev's inequality are the following:

$$\text{Che}(\tau) = \qquad P\{\mu-\tau.\sigma\leq X\leq\mu+\tau.\sigma\}\geq\left(\frac{\tau^2-1}{\tau^2}\right)$$

Che(τ)
0.000
0.750
0.889
0.938
0.960
0.972
0.980
0.984
0.988
0.990
0.992
0.993
0.994
0.995
0.996
0.996
0.997
...

etc

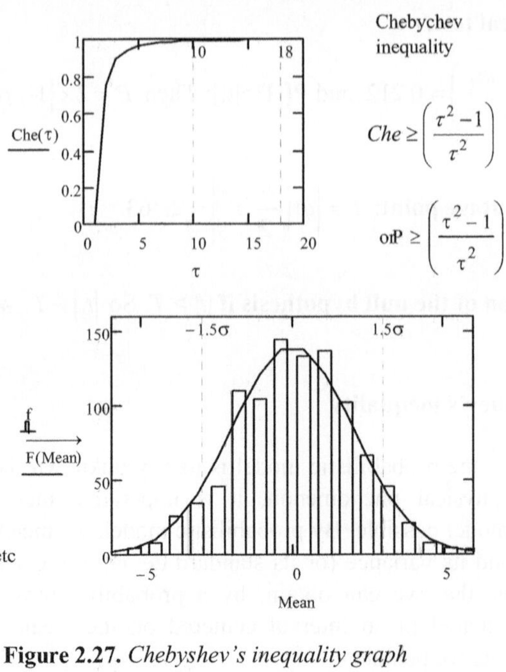

Figure 2.27. *Chebyshev's inequality graph*

The probability that the RV will be situated between ($\mu - 3\sigma$ and $\mu + 3\sigma$) inclusively is at least 0.889. This probability proves to be absolutely important since it does not depend on the probabilistic model equation at all, which as it happens, is unknown. The simple but clever idea that Chebyshev proposes here applies as much to discontinuous RVs as to continuous RVs.

Problem: Sammy received a batch of parts from Nadim who forgot to supply him with the detailed metrological test of the parts. His outdated SPC appliances gave him a mean (μ) of 1.25 inches with a standard deviation (σ) of 0.001 inches. The SPC appliance does not give the distribution but Sammy must specify the centered range surrounding this mean which would include at least 96% of the pieces from the Casanova batch.

Solution: We would use the Chebyshev inequality to solve this problem, as the distribution is unknown to us. Using equation [2.1] we postulate:

$$P\{(\mu-\tau.\sigma)\leq X\leq(\mu+\tau.\sigma)\}\geq\left(\frac{\tau^2-1}{\tau^2}\right),\ \tau\succ 0$$

Answer: for $\mu = 1.25$ and $\sigma = 0.001$ we have $\left(\frac{\tau^2-1}{\tau^2}\right)=0.96\rightarrow\tau=5$ so :

$\{(\mu-\tau\times\sigma)=1.245\ and\ (\mu+\tau\times\sigma)=1.255\}$ According to these data, we will say that at least 96% of the parts have a side situated between 1.245 and 1.255 inches, i.e. a difference of 1/100th inches (0.01 inches).

2.11. Estimation of parameters

It is not sufficient to opt for the application of a distribution to model the physical phenomenon alone. Also, in terms of adequacy tests, it is appropriate to estimate the parameters which characterize the phenomenon observed or materialized to given periods. For this reason there are two possibilities: the punctuality of the value of the chosen parameter and the confidence interval which contains the estimated parameter. Thus, two methods are proposed: the MLE and that of *confidence intervals*. We should point out here that no parameter, however precise, will be able to overcome the uncertainty which mars all the observed phenomena, hence our pedagogical choice regarding uncertainty (Chapter 3).

Practical work: This exercise deals with random numbers generated by the MC method of simulation to generate quantity probabilities whose distributions are unknown. To do this, we will *empirically* determine the probability that the mean of the sample, based on a size = 1,000, of parameters L and s contained in the interval $[a, b]$.

– *Let the parameters be*: $L = 1$, $S = 1$, $a = 0.5$ *and* $b = 1.5$;

– *Let us take the parameters of the MP simulation sampling*:

– *Number of individual samples to collect*: $N_{Samples} = \mathbf{10^4}$;

– *Number of points given in each sample*: $S_{Size} = \mathbf{1{,}000}$;

Let us generate the sampling (N) and the means with the help of Math CAD $i = (0, ..., NSamples-1)$ *and* $Means_i = mean(rlogis(SSize, L,S))$

We can estimate the probability that following relation wants:

$$Success = \sum_{i=1}^{N_{Samples}} \{a \le Means_{i-1} \le b\};\ \text{Prob} = \frac{Success}{N_{Samples}} = 1$$

2.12. Gaussian distribution: estimation and confidence interval

In mechanical reliability engineering, normal distribution is often used to represent component lifespans (mechanisms of usury) at the end of a component's life. This is due to failure rates which are constantly increasing. To do this, we only use it to represent cases whose mean is at least superior to three times its standard deviation.

Practical work: Let us take a crank-connecting rod system with a known Gaussian lifespan with a mean of $\mu = 250{,}000$ hours and a standard deviation of $\sigma = 560$ hours for a task $\tau = 160{,}000$ hours. Calculate and explain the meaning of the results below:

1) The failure rate (before total failure), Z (before 160,000);

2) The estimators of the mean and of the standard deviation;

3) The confidence interval for the mean at the threshold $(1 - \alpha) = 95\%$;

4) Trace the graphs ($f(x)$ and $F(x)$) of the distribution and discuss the limits encompassed by the confidence interval for the mean and the variance (standard deviation).

2.12.1. *Confidence interval estimation for a Gauss distribution (See appendices)*

Mean = μ = 250,000 hour; Standard deviation = 25,000 hour; (Time, task) = 200,000 hour

$$u = \left(\frac{\mu-\tau}{\sigma}\right) = 2 \text{ so } \phi(u) = dt(u,\mu), \text{ also } u = \left(\frac{\mu-\tau}{\sigma}\right) = -2$$

2.12.2. *Reading to help the statistical values tabulated*

$\Phi(u) = 0.053991$

$$f(u) = (1/\sigma)\times\phi(u) \text{ so } f(u) = 2.16\times10^{-6}$$

$\Phi(u) = 1 - \text{pnorm}(\tau,\mu,\sigma)$ so $1 - \text{pnorm}(\tau,\mu,\sigma) = 0.97725$

2.12.3. *Calculations to help the statistical formulae appropriate to normal distribution*

$$\phi(u) = \frac{1}{\sqrt{2\pi}}\times Exp\left\{\frac{-u^2}{2}\right\} \rightarrow \phi(u) = 0.053991 \text{ so } f(u) = 2.16\times10^{-6}$$

While our calculations give $\rightarrow Z(\tau) = 2.209930\times10^{-6}$

2.12.4. *Estimation of the Gaussian mean of unknown variance*

2.12.4.1. *Reliability*

$\boldsymbol{R(\tau)} = 1 - \Phi(u)$ or $R(\tau) = \Phi(u)$ so $\rightarrow R(\tau) = 0.977250$

2.12.4.2. *Failure rate* $Z(\tau)$ of the mechanism*

$$Z(\tau) = \left(\frac{1}{\sigma}\right)\times\phi\left(\frac{\tau-\mu}{\sigma}\right)\Big/\Phi\left(\frac{\tau-\mu}{\sigma}\right) \quad or \quad = Z(\tau) = \frac{f(u)}{\Phi(u)} = 2.21\times10^{-6}$$

data:=

	0
0	$5.000 \cdot 10^4$
1	4.500×10^4
2	4.400×10^4
3	4.000×10^4
4	3.800×10^4
5	3.300×10^4
6	3.600×10^4
7	5.000×10^4
8	4.200×10^4
9	3.900×10^4
10	4.000×10^4
11	3.300×10^4
12	3.500×10^4
13	4.600×10^4
14	2.500×10^4
15	3.200×10^4
16	3.600×10^4
17	5.200×10^4
18	4.400×10^4
19	4.200×10^4
20	3.800×10^4
21	3.700×10^4
22	4.800×10^4
23	4.500×10^4
24	5.600×10^4

Let us enter the vector of repair data predicted for analysis:

n, Number of data to analyze

n: = length(data) alors n = 25

Let us enter a the confidence threshold

$\alpha := 0.05$

Let us calculate the confidence level

So $(1 - \alpha) = 0.95$

Let us calculate the mean, m

$(\mu) = (m = \text{mean(data)})$

So $m = 4.104\times10^4$

Let us calculate the variance, V:

$$V1 := \text{stdev(data)}^2 \cdot \left(\frac{1}{n-1}\right)$$

So $V1 = 2.042 \times 10^6$

Let us calculate the standard deviation, s:

$$s = \sigma = \text{stdev(data)} \cdot \sqrt{\frac{n}{n-1}}$$

s = 1 s

2.12.4.3. *Bilateral confidence intervals for the mean*

$i = 0.n - 1$ with $\alpha = 0.05$

$$\{L\ U\} = \left\{m - \left|qt\left(\frac{\alpha}{2}, n-1\right)\right| \times \frac{s}{\sqrt{n}} m + \left|qt\left(\frac{\alpha}{2}, n-1\right)\right| \times \frac{s}{\sqrt{n}}\right\} \rightarrow$$

Numerical application: $\{L\ U\} = \{3.809 \times 10^4;\ 4.399 \times 10^4\}$ *hours*

Inferior limit L (Lower), and *U*, the *superior limit* for *(Upper)*.

2.12.4.4. *Graphs*

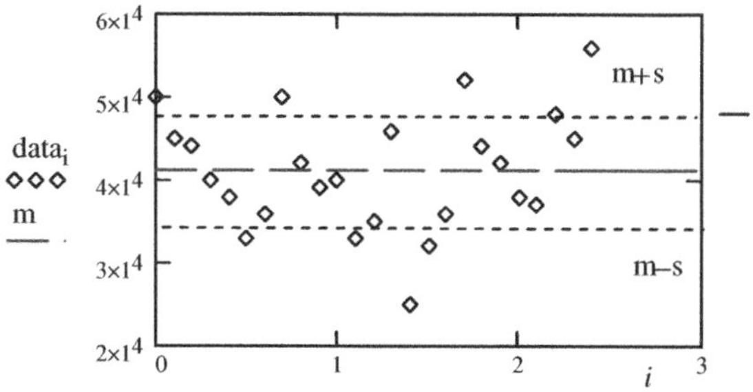

Figure 2.28. *Confidence interval for the mean (m) and the standard deviation (s)*

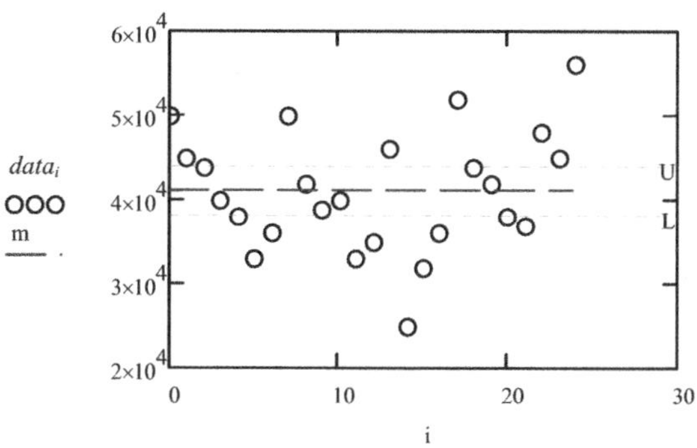

Figure 2.29. *Confidence interval for the mean (m) and the standard deviation (s)*

2.12.4.5. *Bilateral confidence intervals for the variance*

For: $i = (0 \ldots n - 1)$, $\alpha = 0.05$ and $v = (25-1) = 24$, we postulate:

$$\left(\frac{\alpha}{2}\right) = 0.025\ ;\ \left(1 - \frac{\alpha}{2}\right) = 0.975,\ So \rightarrow s = 7.144 \times 10^3$$

See [BOI 01]. From the chi-square statistical tables, we postulate:

$$As\ \chi^2\left(\nu,\frac{\alpha}{2}\right)=\chi^2(24;0.025)=13.8\ we\ postulate:$$

$$Upper\ Limit=\nu\times\frac{(n-1)\times s}{\chi^2\left(\nu,\frac{\alpha}{2}\right)}=\frac{24\times s}{13.8}=1.242\times10^4$$

$$As\ \chi^2\left(\nu,1-\frac{\alpha}{2}\right)=\chi^2(24;0.975)=39.4\ we\ postulate:$$

$$Lower\ Limit=\nu\times\frac{(n-1)\times s}{\chi^2\left(\nu,1-\frac{\alpha}{2}\right)}=\frac{24\times s}{39.4}=4.352\times10^3$$

2.12.4.6. *Bilateral confidence intervals for the standard deviation σ*

$$\sigma_{\prec}=\sqrt{4.325\times10^3}=65.97\ hours \text{ and } \sigma_{\succ}=\sqrt{1.242\times10^4}=111.445\ hours$$

Comments and practical recommendations: It has been shown that the limits of the confidence interval calculated, at the threshold of confidence $\alpha = 0.05$, for the mean, present themselves thus: $\{3.809 \times 10^4 \leq m \leq 3.3999 \times 10^4\}$ hours, i.e. that on average, the lifespan between the 25 parts of the mechanism will be failing on an average going from 3.809×10^4 hours and 4.399×10^4 hours, that is: 5,900 hours.

For the variance: [Inferior limit = 1.242×10^4; Superior limit = 4.352×10^3]

For the standard deviation: $\sigma=\sqrt{s}=84.524 \rightarrow$ we calculate the limits of the differences in this way:

$$\sigma_{\text{inf}}=\sqrt{4.352\times10^3}=65.97\ hours\ and\ \sigma_{\text{sup}}=\sqrt{1.242\times10^4}=111.445\ hours$$

2.13. Kaplan–Meier estimator

There are different ways of choosing the distribution model which would best convey the theory/practice concordance to give a certain and true indication reliability. In the domain of mechanics of continuous zones which interests us in this text, we will often focus on three of four models, which are:

– *the extreme value model*;

– *the multiplicative degradation model*;

– *the lifespan in fatigue represented by the Birnbaum-Saunders model.*

It is important to remember that the choice of distribution law of lifespan is essentially based on the appropriate failure trend. The Kaplan-Meier technique can be applied without making assumptions on the lifespan distribution law. However, it is imperative to have abundantly supplied data. Acceleration modeling proves to be difficult to put into practice. Without any known or assumed distribution, the KM approach [KAP 68] can be applied to any level (mode, component, system, structure etc.). Known as a *product-limit estimator*, it allows estimation of the survival function. This property allows the tracing of an ensemble of events relative to total failure or partial failure of a structure within a given time. An important advantage of the Kaplan-Meier curve is that the method can take "censored" data into account. When no truncation or censoring occurs, the Kaplan-Meier curve is equivalent to empirical distribution.

Formulation of the KM estimator: Let us take $P(\tau)$, the probability that a component of a given structure will have exceeded the lifespan (τ). For a sample coming from a population of size N, the time observed as far as failure (total failure) of N such as the correspondence of each of the τ_i is n_i, the number "subject to failure" just before time τ_i. d_i is the number of total failures with time τ_i. We can see that the intervals between each occurrence will not be uniform. For example, a structure could begin with x failing components on day j (censored test) and with a different failure on day $(j + 1)$. The KM estimator is *non-parametric (free distribution).* It is in fact, an evaluation of maximum likelihood. More *robust* in reliability, non-parametric methods have more widespread applications than parametric methods.

When the test is not censored, n_i is the number of survivors just before time τ_i.

For the censored test, n_i will be the number of survivors minus the number of loss. It is only these cases of survival which are always observed (not yet censored) which are threatened by failure (observed). *Non-parametric models* are *deductive statistical methods.*

2.13.1. *Empirical model using the Kaplan–Meier approach*

The KM evaluation is an empirical procedure (non-parametric). The KM procedure gives fast and simple evaluations of the reliability function. It is based on failure data which can be multi-censored. No underlying model (such as Weibull or

Log-Normal) is assumed. The exact time of failure is demanded however. The stages to follow can be summarized as:

– ordering real failure times from τ_1 to τ_r, where there are (r) failures;

– associating the number n_i, with n_i at each τ_i,= the number of units of exploitation just before the i th failure happens in time τ_i;

– evaluating $R(\tau_1)$ by $R(\tau_1) = \{(n_1 - 1)/n_1\}$;

– evaluating $R(\tau_i)$ by $R(\tau_i) = R(\tau_{i-1}) \times [(n_i - 1)/n_i]$;

– evaluating the repair function $F(\tau_i)$ by $(1 - R(\tau_i))$.

We do not consider the data of the components which are not failing (censored test). We only consider them at the *last* moment of failure before their withdrawal. They are included in the n_i counting during the time-to-fail, but not after.

2.13.2. *General expression of the KM estimator*

Let us hypothesize n components from a batch to be tested and sort the observed times for these n components from τ_1 to τ_n. Some within this group are real times to fail and some are durations for test units taken before they fail. This is following the trace of all the indices corresponding to the instantaneous times of failure. We present the KM estimators below:

$$\hat{R}(\tau_i) = \prod_{\substack{j \in S \\ \tau_j \le \tau_i}} \left(\frac{n-j}{n-j+1} \right) \qquad [2.87]$$

– A light modification of the KM estimator produces the best probability results. $\hat{R}(\tau_i)$ is a corrected estimator and S is a play on all the indices (j) just as τ_j is an instantaneous time-to-fail. The notation $J \in S$ and $\tau_j \le$ to τ_j signifies that we manipulate the products of the j indices contained in S corresponding to failure times $\le$ to τ_j.

– Once reliability $R(\tau_{j.})$ is calculated, we will evaluate the cumulative distribution function of probabilities using the connection: $F(\tau_{j.}) = 1 - R(\tau_{j.})$.

Modified KM estimator: At time τ of the last failure, the KM estimator has a reliability of $R(\tau_j) = 0$ and $F(\tau_{j.}) = 1$. This estimator has a bias that is subject to caution, as the cumulative distribution function gets *asymptotically* closer from 1 to time τ which leans toward the infinite. The modified KM estimator is written:

$$\hat{R}(\tau_i) = \left(\frac{n + \frac{7}{10}}{n + \frac{4}{10}} \right) \times \prod_{\substack{j \in S \\ \tau_j \leq \tau_i}} \left(\frac{n - j + \frac{7}{10}}{n - j + 1.7} \right) \qquad [2.88]$$

2.13.3. *Application of the ordinary and modified Kaplan–Meier estimator*

1) We tested a batch of $N = 57$ small connecting rods in good working order. Seven (7) of them suffered failures which came about during processes of normal functioning with times τ_i given as, *f{*$\tau_1 = 13$; $\tau_2 = 20$; $\tau_3 = 56$; $\tau_4 = 77$; $\tau_2 = 200$; $\tau_3 = 360$; $\tau_4 = 417$*} hours.*

2) At regular intervals (10 hours), we subtracted, in a test, six small connecting rods after *{62; 72, 82, 92, 102 and 112} hours.*

3) The 44 small connecting rods will be objects of ulterior examination as they are instantaneously non-failing. They will be withdrawn after 3 months of normal functioning at the rate of two *times* seven *hours* per day team.

$$\left\{ \begin{array}{l}
R(\tau_1 = 13) = \left(\frac{56}{57}\right) = \\
R(\tau_2 = 20) = \left(\frac{56}{57}\right) \times \left(\frac{55}{56}\right) = \\
R(\tau_3 = 56) = \left(\frac{56}{57}\right) \times \left(\frac{55}{56}\right) \times \left(\frac{53}{54}\right) = \\
R(\tau_4 = 77) = \left(\frac{56}{57}\right) \times \left(\frac{55}{56}\right) \times \left(\frac{53}{54}\right) \times \left(\frac{52}{53}\right) = \\
R(\tau_5 = 200) = \left(\frac{56}{57}\right) \times \left(\frac{55}{56}\right) \times \left(\frac{53}{54}\right) \times \left(\frac{52}{53}\right) \times \left(\frac{50}{51}\right) = \\
R(\tau_6 = 360) = \left(\frac{56}{57}\right) \times \left(\frac{55}{56}\right) \times \left(\frac{53}{54}\right) \times \left(\frac{52}{53}\right) \times \left(\frac{50}{51}\right) \times \left(\frac{47}{48}\right) = \\
R(\tau_7 = 417) = \left(\frac{56}{57}\right) \times \left(\frac{55}{56}\right) \times \left(\frac{53}{54}\right) \times \left(\frac{52}{53}\right) \times \left(\frac{50}{51}\right) \times \left(\frac{47}{48}\right) \times \left(\frac{44}{45}\right) =
\end{array} \right\} \text{etc. } \ldots$$

2.14. Case study of an interpolation using the bi-dimensional *spline* function

This simulated example shows the mathematical reasoning which allows the conduction of the 2D interpolation of a surface, with the help of *cspline* and *interp* functions from the MathCAD program. For this, we will enter a matrix defining a surface where the number of lines of the matrix equals the number of columns.

$$\text{Mintspline} = \begin{pmatrix} 0.125 & 0.141 & -0.235 & -0.417 & -0.285 & 0.327 & 0.333 \\ 0.999 & 0.117 & -0.26 & -0.986 & -0.312 & 0.826 & 0.253 \\ -0.349 & -0.334 & -0.44 & 0.24 & 0.215 & 0.305 & -0.225 \\ -0.553 & -0.225 & 0.425 & 0.336 & -0.106 & 1\times 10^{-3} & 0.117 \\ -0.979 & 0.175 & 0.368 & 0.314 & 0.355 & -0.66 & 0.203 \\ -0.707 & 0.125 & 0.361 & 0.302 & 0.303 & -0.164 & 0.204 \\ 0.525 & 0.118 & 0.566 & 0.128 & 0.133 & -0.125 & 0.222 \end{pmatrix}$$

We state the vector n, X and Y to determine the linking of the matrix

$$\begin{pmatrix} lines(\text{mintspline}) = 7 \\ cols(\text{mintspline}) = 7 \\ n = lines(\text{mintspline}) \end{pmatrix} \rightarrow \left\{ X = \begin{pmatrix} 0 \\ 1 \\ 2 \\ 3 \\ 4 \\ 5 \\ 6 \end{pmatrix} \text{ and } Y = \begin{pmatrix} 0 \\ 1 \\ 2 \\ 3 \\ 4 \\ 5 \\ 6 \end{pmatrix} \right\}$$

The vectors X and Y used are: Mxy: = increase(tri (X), sort (Y)); lines (Mxy) = 7

The calculated spline coefficients are: Cef := cspline(Mxy, Mintspline)

Surface adjustment function: $fit(x,y) = \text{interp}\left\{ coef, Mxy, \text{Mintspline}, \begin{pmatrix} x \\ y \end{pmatrix} \right\}$

Examples of interpolated values: fit (2.5, 3.9) = 0.077 and fit (0.1, 1.7) = −0.025

$$\text{xlow} := \text{Mxy}_{0,0} \quad \text{xhigh} := \text{Mxy}_{n-1,0}$$
$$\text{ylow} := \text{Mxy}_{0,1} \quad \text{yhigh} := \text{Mxy}_{n-1,1}$$

Density of linkage for interpolation: *xn* = *4xn* and *yn* = *4xn*

$$fit(x,y) = \text{interp}\left\{coef, Mxy, \text{Mintspline}, \begin{pmatrix} x \\ y \end{pmatrix}\right\}$$

$$\text{For i:= 0..xn} - 1 \quad \text{xind}_i := \text{xlow} + \text{i} \cdot \frac{\text{xhigh} - \text{xlow}}{\text{xn} - 1}$$

$$\text{For j:= 0..yn} - 1 \quad \text{yind}_j := \text{ylow} + \text{j}.\frac{\text{yhigh} - \text{ylow}}{\text{yn} - 1} \quad \text{FIT}_{i,j} := \text{fit}(\text{xind}_i, \text{yind}_j)$$

3D graph

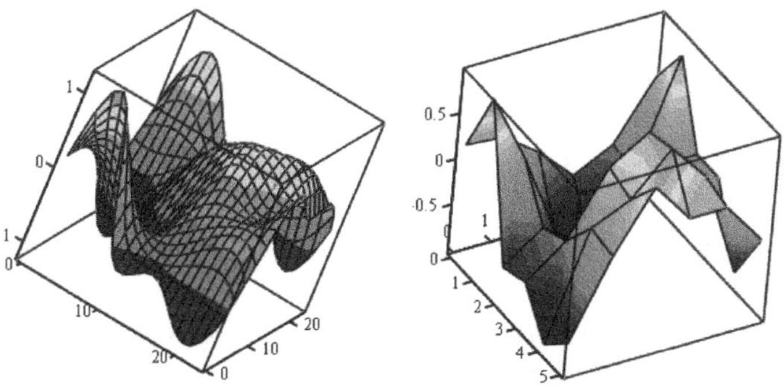

FIT Mz

Figure 2.30. *Illustration of a 2D interpolation of a surface (with the help of MathCAD) FIT = Interpolated spline surface 2D spline (2D) and Mz = Minspline or original surface*

2.15. Conclusion

The characterization of probability laws with uncertain variables is, in itself a sort of risk mitigation. This is true due to the knowledge of the hazard which results from it, and as a result, its acknowledgement comes across. Knowing uncertainty is already a way of reducing (even attenuating) risk or at any rate trying to better apprehend it. The statistical tools are made use of to this end. Adequacy and adjustment tests remain very useful in the applicability of distribution laws for the modeling of the lifespan of materials and of structures.

2.16. Bibliography

[BIR 52] BIRNBAUM Z.W., "Numerical tabulation bof the distribution of Kolmogorov's statistic for finite sample size", *Journal of the American Statistical Association*, vol. 47, pp. 425–441, 1952.

[BOI 01] BOITEUX B., "Fractiles de la loì du khi-deuk", *Guide du technicien qualité*, Éditions Delagrave Paris, France, 2001.

[BOX 74] BOX G.E.P., MCGREGOR J.F., "The analysis of closed-loop dynamic stochastic systems", *Technometrics*, vol. 16, no. 3, 1974.

[BOX 94] BOX G.E.P., JENKINS G.M., REINSEL G.C., *Time Series Analysis, Forecasting and Control*, 3rd ed., Prentice Hall, Englewood Clifs, NJ, 1994.

[CHA 67] CHAKRAVARTI L., ROY L., *Handbook of Methods of Applied Statistics*, vol. I, p. 392–394, John Wiley & Sons, New York, 1967.

[DAT 10] LOGICIEL du gouvernement des USA-NIST Utilisation libre, http://www.itl.nist.gov.

[DUA 64] DUANE J.T., Learning Curve Approach to Reliability Monitoring, *IEEE Transactions On Aerospace*, 2, pp. 563–566. USA, 1964.

[D'AGO 86] D'AGOSTINO R.B., STEPHEN M.A., *Goodness-of-Fit Techniques*, Marcel Dekker, Inc., New York, 1986.

[FIS 61] FISK P.R., "The graduation of income distributions", *Econometrica*, vol. 29, pp. 171–185, 1961.

[GRO 94] GROUS A., Etude probabiliste du comportement des Matériaux et structure d'un joint en croix soudé, PhD thesis, UHA France, 1994.

[GRO 98] GROUS A., RECHO N., LASSEN T., LIEURADE H.P., Caractéristiques mécaniques de fissuration et défaut initial dans les soudures d'angles en fonction du procédé de soudage, *Revue Mécanique Industrielle et Matériaux*, vol. 51, n° 1, Paris, April 1998.

[GRO 09] GROUS A., *Métrologie appliquée aux sciences et technologies*, Hermès, Hermes-Science/ Lavoisier, France 2009.

[GRO 11] GROUS A., *Applied Metrology for Manufacturing Engineering*, ISTE Ltd, London, John Wiley & Sons, New York, 2011.

[KAP 58] KAPLAN E.L., MEIER P., Evaluation non paramétrique à partir des observations inachevées, *Journal. AM. Statut. Association*, 53, 457–481, 1958.

[MON 91] MONTGOMERY D.C., *Design and Analysis of Experiments*, John Wiley & Sons, New York, 1991.

[MOO 86] MOORE M.K., MOORE D.S., "Tests of chi-square type", in D'AGOSTINO R.B., STEPHENS M.A. (eds), *From Goodness-of-Fit Techniques*, Marcel Dekker, New York, 63–95, 1986.

[SHA 65] SHAPIRO S.S., AND WILK M.B., An analysis of variance test for normality (complete samples), *Biometrika*, 52, 3 and 4, pages 591-611, 1965.

[SNE 89] SNEDECOR G.W., COCHRAN W.G., *Statistics Methods*, 8th ed., IOWA State University Press, 1989.

[STE 74] STEPHEN M.A., "EDF statistics for goodness of fit and some comparisons", *Journal of the American Statistical Association*, vol. 69, pp. 730–737, 1974.

[STE 76] STEPHEN M.A. "Asymptotic results for goodness-of-fit statistics with unknown parameters", *Annals of Statistics*, vol. 4, pp. 357–369, 1976.

[STE 77] STEPHEN M.A., "Goodness of fit for the extreme value distribution", *Biometrika*, vol. 64, pp. 583–588, 1977.

[STE 79] STEPHEN M.A., "Tests of fit for the logistic distribution based on the empirical distribution function", *Biometrika*, vol. 66, pp. 591–595, 1979.

Chapter 3

Modeling Uncertainty

3.1. Introduction to errors and uncertainty[1]

There are no measurements which are *absolutely* exact. Measurements are always tainted with imputable errors due to various human and material causes. Qualifying an error to qualify an element of uncertainty proves what we have doubts about the validity of the result of this measurement. Evaluating elements of uncertainty based on measurements that introduce errors will have negative consequences. "Signposting" such influencing factors depends on the type of measurement used and we will begin by developing the mathematical principles that are suitable for this domain [NIS 94, NIS 97]. Meanwhile, we will discuss, among others, three major sources of uncertainty:

– environmental elements (wind, running expenses);

– geometrical and physical data;

– the imperfect character of theoretical models used to represent phenomena (idealizations, imperfections, erroneous comprehension of problems, etc.).

These uncertainties can be classified into two important categories:

---

1 This chapter constitutes a broad summary of the volumes already published in the [GRO 09, GRO 11]. We have essentially kept the elements of analysis dedicated to the type A and type B uncertainty to apply them to mechanical reliability of materials and structures and to quality control.

Physical uncertainties *aka* intrinsic uncertainties: they arise from the characteristics of a physically random phenomenon (speed and orientation of the effects of the wind, the waves and the emergence of an ice storm). These uncertainties cannot be eliminated or even reduced.

Uncertainties of knowledge: these are the measurement uncertainties which are due to the inherent imperfections of the instrument in use. They represent statistical uncertainty due to a lack of information.

Many laboratories participate in interlaboratory studies where the method of testing is evaluated in terms of its *repeatability* and its *reproducibility*. Uncertainty analysis aims to evaluate the result of *particular measuring*, in a *special test laboratory* at a *given moment*. The two objectives are linked. If a test laboratory has taken part in an interlaboratory test which follows recommendations and material analysis [EHR 98] of a standard test (ASTM E691) or a norm [ISO 97b], it can represent its *standard uncertainty* for a single measure like the standard deviation of *reproducibility*. When reading [VIM 93] the *International Vocabulary of Metrology* (VIM) and the *Guide to the Expression of Uncertainty in Measurement* (GUM), and section 3.7.2 of ISO 1087-1, 2000 norm, the definition assigned to words in *metrology* such as *error and uncertainty* is often truncated.

The development of the treatment of measurement uncertainty using the classical approach or the *true value* (*forever unknown*) toward an uncertainty approach has led to a reconsideration of certain concepts. We admit that instruments and measurements do not allow for this *true value*.

We understand that it is possible to distinguish between these two categories of error, treated differently in the *propagation of errors*. However, no justified rule has come forward to combine systematic errors and additive random errors to make a total error, thus characterizing the result of this measurement. Despite this, it is possible to estimate a superior limit of total error *clumsily* named uncertainty. The components of measurement uncertainty [VIM 93] are conventionally grouped into two categories. The first, named *type A*, is estimated by *statistical methods*. The *type B* category is primarily based on distributions. It is, in fact, the operator who must evaluate the sources of error. The manufacturer supplies data as an *application class*, *measurement standards*, *resolution*, etc. In combining the two categories, A and B, we obtain composed uncertainties $U_c(y)$.

3.2. Definition of uncertainties and errors as in the ISO norm

Uncertainty, according to the ISO guide (GUM) and the VIM is a "parameter, associated with the follow-up of a measurement, which characterizes the dispersion of values which could reasonably be attributed to the measurand". The reference values will be the consensual values based on measurements taken according to a specific protocol by a group of laboratories. The components of uncertainty are merged to classify the sources of error into two large categories, according to the source of the data and not according to the type of error. Each component is quantified by a *standard deviation* whose categories are of *Type A* (*statistical methods*) or of *Type B* where all the components are evaluated by *other means* (*or from other laboratories*). If the object of the declaration of uncertainty is to supply a coverage with a raised level of confidence, increased uncertainty will be calculated as follows:

$$U = \kappa \times u \qquad [3.1]$$

where κ is chosen to be the $\tau_{\alpha/2,(\nu)}$ critical value of Student's t-table for v degrees of freedom (*df*) at a confidence threshold of 5%. For large *df*, we use $\kappa = 2$ to approximately 95% of coverage (expansion). The *df* for *type A* uncertainties are the *df* for their respective standard deviations. The *df* for the *type B* evaluations are obtained from or are certificates of the measurement standardization of published reports. Particular cases where the standard deviation must be estimated based on fragmentary data or scientific judgment are estimated as having infinite df; for example:

– mini-estimation based on a study of robustness and other tests;

– estimation based on presumed cumulative distribution of possible errors;

– uncertainty component of type B for which the df (ν) are not tabulated.

The *df* (ν) for standard uncertainty (u), which can be a combination of numerous standard deviations, are not generally known. This is irritating if there are large uncertainty components with small df. In this case, the df are approached using the Welch–Satterthwaite [BRO 60] formula.

$$\nu = \frac{(u^4)}{\sum_{i=1}^{R} \frac{\psi_i^4 \times \sigma_i^4}{\upsilon_i}} \qquad [3.2]$$

where:

u is standard uncertainty;

σ_i is the standard deviation;

Ψ_i are the sensitivity coefficients;

υ_i are the small *df.*

Random error and bias error take the estimation of possible divergence and errors during the measurement process into account. These errors can be corrected or eliminated. Precision and bias are dependent on the measurement method. Uncertainty is thus a property of a *specific result* for an element of unique testing which depends upon a configuration of specific measurements. That depends on the *repeatability* of an instrument, on the *reproducibility* of the result over time, on the number of measurements in the test result, and on the sources of random and systematic error which could contribute to a disaccord between the result and its reference value. The VIM terms encompass the terminology rules shown in the ISO norms: ISO 704, ISO 1087-1, and ISO 10241. Uncertainty makes us aware of the way in which we measure a value. It changes the *confidence* accorded to a result which implies standardizations and appropriate calculations. A measurement standard is defined as:

> "The materialized measurement using an measurement device or a system of measurement destined to define, produce, conserve or reproduce a unity or several known values of a certain size to transmit them in comparison to other instruments of measure".

– *First test standard*: it is a "standard which is designed or largely recognized as presenting the highest metrological qualities and whose value is established without referring to other measurement standards of the same size".

– *Reference test standard*: it is a "measurement standard, in general, of the highest metrological quality available in a given place or in a given organization, from which measurements based on it come from".

– Transfer test standard: it is a "measurement standard used as an intermediary to compare standards to each other".

– *Work test standard*: it is a "standard which is commonly used to calibrate or test materialized measurements, measurement instruments or reference materials".

Measuring is the *set of operations having the determination of the value of a size as its purpose.* The size submissive to measuring is the *measurand.* The evaluation of a size in comparison with another size of the same type taken as a unit gives rise to measure *X.* The size of *X* is a parameter tested during the elaboration of a product or of its transfer. The calibration certificate of an instrument [CLA 00] gives a difference and an uncertainty on this difference, which we call a *calibration uncertainty.* Measurement uncertainty encompasses the following uncertainties:

– measurement calibration which happens during joining;

– due to the accuracy of an instrument if uncorrected;

– linked to the derivative (fatigue, etc.) of the instrument between two calibrations;

– linked to the instrument's own characteristics (readings, repeatability, etc.);

– linked to the environment, if the conditions differ from those of the calibration.

According to the VIM, measurement standards are defined as "materialized measurements, measuring devices or measuring systems destined to define, produce, conserve or reproduce a unit or several known values of a size to transmit them by comparison to other measuring instruments".

By analogy of what precedes, the definitions of the four main measurement standards are as follows:

– *International standard*: this is a standard recognized by an international agreement to serve as an international foundation for the fixation of values of all the other measurement standards of size concerned.

– *National standard*: this is a measurement standard recognized by an official national decision to serve as a foundation in a country for the fixation of values of all the other measurement standards of size concerned.

– *Primary standard*: this is a standard which presents the highest metrological qualities in a specified domain. It is therefore a standard with the highest order of precision in a measurement standardization system, which is used to calibrate a standard of the lowest level.

– *Secondary standard*: this is a designed element of measurement, used in a measurement calibration system as a means of transfer of a basic value from a primary standard, to measure another piece of test equipment or element.

Traceability supports ordered and permanent filing which allows the user to get to know the history of a process or of an instrument, to know the derivative or the evolution of a material. According to the VIM, traceability "is the property of the result of a measure or the value of a measurement standard by which it can be linked to specified references (national standards or international standards), by an uninterrupted chain of comparisons having clarified all uncertainties".

3.3. Definition of errors and uncertainty in metrology

Uncertainty is *estimation* characterizing the range of values in which the *true value* of a *measured size* is situated. In fact, measurement uncertainty is made up of several components. Some may be estimated with the help of the statistical distribution of a series of measurements (experimental standard deviation). The estimation of other components can only be based on experience.

3.3.1. *Difference between error and uncertainty*

The following illustration presents the error which is included in the interval which separates the measured value from the true value. Uncertainty has never signified error in metrology [VIM 93, GUI 04].

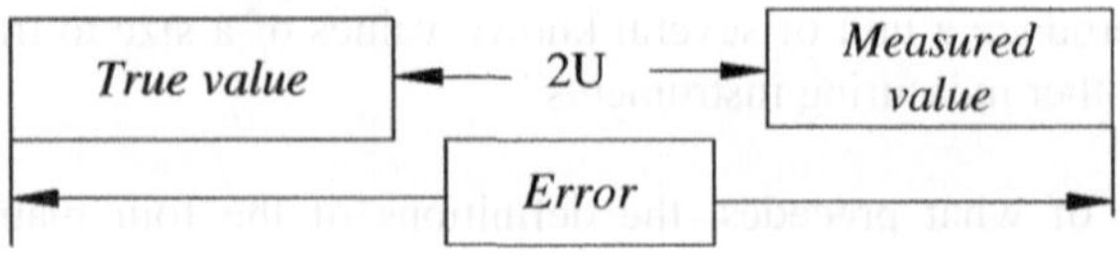

If T is the tolerance admitted on a measure, we postulate the uncertainty expression:

$U = \{-T \prec error \prec +T\}$, so the result will be equal to the measured value $\pm U$.

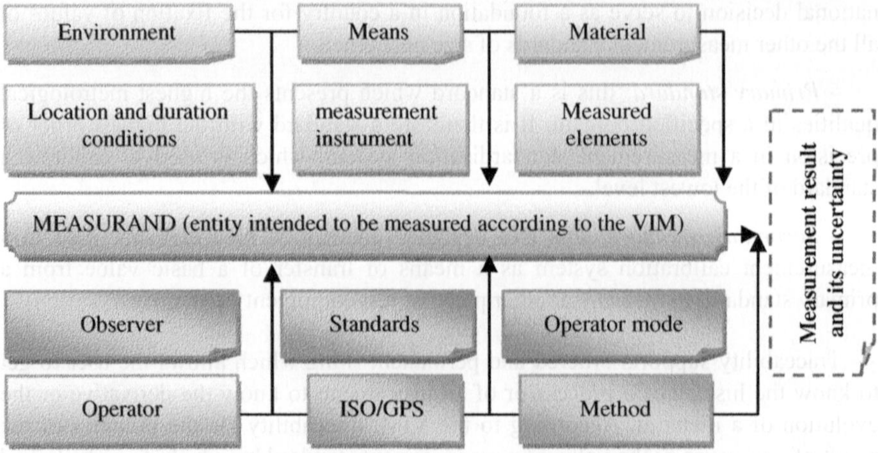

Figure 3.1. *Illustration of error compared to uncertainty (VIM, International Vocabulary of Metrology)*

We distinguish between two main types of error whose measurement may be affected. These are *systematic error* and *accidental error*.

3.3.1.1. *Accidental or fortuitous errors*

These are the result of an incorrect maneuver, of incorrect usage, or of instrument dysfunction. We do not know how to quantify them without adding to error itself. These errors cannot be avoided as their cause is due to human error. When we measure it, the result x of size X is not totally defined by a single number, the uncertainties come from different errors linked to measurement. It is necessary to at least characterize it using a pair $(x, \delta x)$ and a unit of measurement $\{x - \delta x \leq X \leq x + \delta x\}$. We think that these errors, linked to the entity measured and to the system of observation, do not automatically permit the deduction of the true value.

$$\textit{Result of the measurement} = \textit{true value} + \textit{error} \quad [3.3]$$

3.3.1.2. *Systematic errors*

These are reproducible errors linked to their cause by a physical law, therefore susceptible to being eliminated by suitable corrections. According to the VIM, systematic error is defined as the "mean which results from an infinite number of measurements of the same measurand in conditions of repeatability, minus a true value of the measurand".

$$\textit{Systematic error} = \textit{error} - \textit{random error} \quad [3.4]$$

Like the true value, systematic error and its causes cannot be completely known. Being a question of an instrument of measure, we will refer to the definition of *error of accuracy* translated as follows:

$$\textit{Result} = \textit{true value} + \textit{random error} + \textit{systematic error} \quad [3.5]$$

Systematic errors occur during the employment of badly calibrated units on an erroneous scale. The main characteristic of these defects is that they always act in the same way on the result of the measurement, distorting it systematically using excess or defect. We say that these defects introduce systematic errors know as defects of *precision or of accuracy* [REE 67]. The attitudes to adopt based on the above are listed as follows:

– being conscience of the existence of errors, being aware of the existence of errors, and above all, avoiding obscuring them;

– ensuring to detect them knowing that they always act in a determined way;

– reducing their influence by ultilizing the instrument in an appropriate manner (adjusting to zero);

– if that is impossible, carrying out a correction which takes signaled defects into account.

3.3.1.3. *Errors due to instrument*

Errors imputable to measuring instruments [CAT 00] are often inherent to mechanical defaults. The accuracy of a measurement instrument is defined by the interval of its measurement scaling.

3.3.1.4. *Errors due to the operator*

Sometimes, measurement errors which come from an evaluation are erroneous or tainted with imprecision. Traits not coinciding with scaling and the inaccurate visual position affect precision and therefore judgment.

3.3.1.5. *Errors due to temperature differences*

Hygrometric conditions influence the properties of materials (instruments, devices, and parts subject to measuring). Temperature differences have an influence as much on measurement instruments as on the components to measure and particularly on instruments which have been meticulously adjusted. Measurement instruments are calibrated at 20 °C. For the initial length L and a coefficient of dilation of materials α, $\Delta\tau = (\tau_1 - \tau)$ is the variation of temperature in °C. Lengthening is expressed by the following relation:

$$\Delta L = L \times \lambda \times (\tau_1 - \tau) \quad [3.6]$$

It would be utopian to believe that all measurements refer *truly* to what is produced by an instrument, however precise it may be. This implies that, any measurement operation would be tainted with errors or uncertainty. The origins of these different errors are sometimes difficult to distinguish. However, metrologists agree upon classifying them into three categories: errors due to *values themselves,* due to the *measurement system*, and the *measuring method.* We must clearly define values in a precise manner, as an erroneous definition of this would inevitably distort interpretation. In practice, the perfect measurement system does not exist. When we repeat the same measurement several times, dispersions occur and prove this methodological fact. Sometimes, measurement standards, having survived calibration, are inaccurate. According to the VIM, random error is a "result of measuring minus the mean of an infinite number of measuring of the same measurand, carried out in conditions of repeatability".

$$\textit{\textbf{Random Error}} = \textit{error} - \textit{systematic error} \quad [3.7]$$

Random errors are non-repeatable. Let us consider a size X to be measured several times, in identical conditions with measurements which are independent of each other. In spite of these precautions, we end up getting results which are different from one another. This is where *reliability* defects appear. We can see them via the non-repeatability of the results. These defects contain *random errors,* hence the statistical treatment of the results to achieve an estimation of a given uncertainty. It would be misleading to try and use an instrument beyond the limits of its sensitivity. First, we must determine precision and a method appropriate to the purpose, thus envisaging the possibility of instrumental and methodological accuracy defects. We then make the necessary adjustments to minimize adjustment defects. Most characteristics come into play in uncertainty evaluation such as repeatability, reproducibility, linearity, sensitivity, precision, resolution, reliability, adequacy, and accuracy.

3.3.1.5.1. Repeatability

According to ISO (3534-1 and 5725-1), repeatability is the minimal value of reliability. In fact, it is the dispersion of independent measurements obtained based on identical samples by the same operator, and using the same equipment in a short interval of time. Repeatability is evaluated in the domain studied at κ level of concentration while repeating n measurements for each sample. The standard deviation σ_r is as follows:

$$\sigma_r = \sqrt{\frac{\sum_{i=1}^{n}(x_i - \bar{x})^2}{n-1}} \qquad [3.8]$$

3.3.1.5.2. Reproducibility (internal)

At least one of the factors varies in comparison with repeatability: the time factor or the operator factor. We consider the effect of the factor studied, using the analysis of the variance σ^2_r. By repeating the same measurement experiments, we highlight variability, hence approximation using the expression of variance $(\sigma_r)^2$.

Scale error: this is an error which depends on the linear method of the measured size. With time, aging makes the components tighten toward what is usually called the "drift". In addition to the effect of sizes of influence, in sensor metrology, we must appreciate the aging of components, by expressing the variation of their output signals.

Linearity error: linearity is used to express the linear relationship between the results obtained in all measurement areas and the properties corresponding to the material. A nonlinear relationship is generally eliminated using correction by means of the function of nonlinear measurement standards. Mobility error has the characteristic of being step shaped, due to a digitization of the signal.

Error due to the hysteresis phenomenon: there is a hysteresis phenomenon when the result of a measurement depends on *the* conditions of use *earlier in* the previous measurement. It is reversibility which characterizes the aptitude of an instrument to give the same indication when we achieve the same value of size measured by increasing values or by decreasing values.

We define the *range* as the "collection of values of the measurand for which the error of an instrument of measure is supposed to be comprised between specific limits. The maximal value of the measurement range is called full-scale". The calibration curve is specific to each instrument. It allows the transformation of the gross measurement into a corrected measurement. It is obtained by referring the instrument to a true value of the size to measure, supplied by a measurement standard, and by reading the given gross measurement with precision.

3.3.1.5.3. Sensitivity

This is the result of the increase in the response of a measurement instrument by the increase corresponding to the input signal. This definition from the VIM [VIM 93] applies to instruments and devices with various signals. In other words, it is a parameter which expresses the variation of the exit signal of a measuring device (instrument) according to the variation of the input signal. Let us see how this is presented:

Let X be the size to be measured and x the signal supplied by the instrument. All of the values of X lie within the measurement range corresponding to a value of x such that f(X) = x. The sensitivity surrounding a value of X is q such that q = dx/DX. If the function is linear, the sensibility of the instrument is constant such that q = Δx/ΔX.

Sensitivity expresses the smallest quantity Δx measured for a determined value x of a measured size. This sensitivity can be constant throughout the scale. It is even larger when the number dx is small. When x and X are similar, q is the gain in decibels. If I is the signal given by the test, and α the number of variables to be measured, then S will be the sensitivity in the vicinity of a given value U of the size of sensitivity α to be measured. We must consider the gradient of the scale curve on a certain interval, to present the mean sensitivity:

$$\alpha_{mean} = S = \left(\delta U/\Delta\alpha\right) \quad [3.9]$$

EXAMPLE.–

$$\left\{\begin{array}{l} For\ \alpha = (1\ldots10), f(\alpha) = 20.Log(\alpha) the\ gain\ is\ written\ f(\alpha) = 20.Log(\alpha)\ (dB) \\ U = 1\ ;\ \alpha = (1\ldots10), f(\alpha) = U \times \alpha \rightarrow \alpha = \dfrac{\Delta U}{\Delta U} = \text{Arctan}(\varepsilon) = the\ slope\ (gradient) \end{array}\right\}$$

It is important not to confuse sensitivity with resolution. Resolution is the smallest variation in size to be measured so that the instrument is capable of detecting. For example, a dial test indicator indicates 100 mm. A variation of 0.1 mm makes the dial's needle move while a variation of 0.05 mm does not make the needle move. The resolution of the indicator is thus 0.1 mm. When precision is limited by the instrument's sensitivity, we accept that all measurements X give an estimation which is sufficient to the true value of size X and we are simply looking to protect ourselves from gross errors (measurement errors) by carrying out two successive measurements X_1 and X_2.

d(q) =

	0
0	0.000
1	6.021
2	9.542
3	12.041
4	13.979
5	15.563
6	16.902
7	18.062
8	19.085
9	20.000

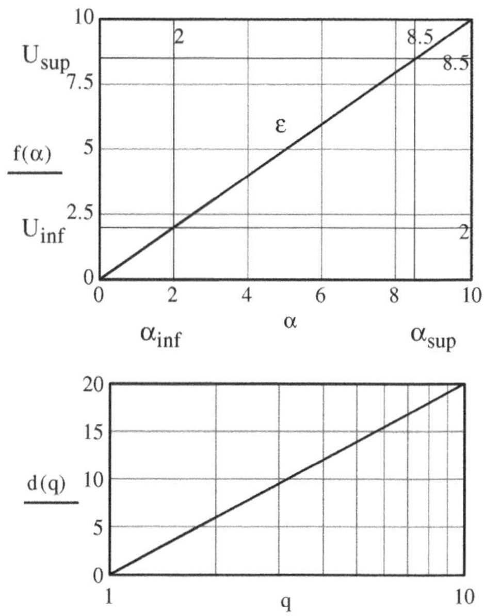

Figure 3.2. *Examples of determination of sensitivity and gain line*

Let us consider R as the resolution of the instrument. We consider the results to be concordant if $|X_1 - X_2| < R$. We will adopt the following as measurement results: $X = (x_1 + x_2)/2$ and we will write it: $\{x - R \leq X \leq x + R\}$. The relative uncertainty will be (R/X) and the precision of the measurement is written in as $(R/X)\cdot 100$ in %. If the two successive results x_1 and x_2 do not tally, it will be necessary to carry out a third test x_3. If x_3 agrees with one of the two test parameters, for example x_1, this will lead us to reject x_2 for being aberrant and finally give the following as the result of the measurement: $X = (x_1 + x_3)/2$, etc. So if x_3 does not agree with any of the first two results, that could be a sign of limited precision, not due to the sensitivity of the instrument, but rather due to reliability defects. *Reliability defects limit the precision of the measurement.*

Resolution targets digital display instruments to express the smallest value displayed. *Robustness* expresses resistance or insensitivity to the effects of certain sizes of influence. *Selectivity* is the capacity to correctly carry out a measurement in spite of the influence of *interferences*, for example the capacity of the method to distinguish two objects which are close to each other in accuracy (qualitative).

3.3.1.5.4. Precision

Precision expresses the degree of concordance of independent measurement characteristics resulting from the application of a process under specified conditions. When we talk about a class of precision, we target the class of a measuring instrument which corresponds to the value in percentage of the relationship between the largest possible errors on the measurement range. A measuring instrument is characterized by at least one number, called a class indicator. Precision is easier to define using precision error. It is expressed in size units (absolute error) or in percentage (relative error). Apart from operating conditions, the precision of an instrument is essentially linked to two types of characteristics: accuracy and reliability. An instrument is qualified as being precise if it is both accurate and reliable at the same time.

In practice, precision is an element which globally fixes maximum error (plus or minus) being able to be implemented during measurement. It is generally expressed in percentage of the measurement range. At these maximal values of scale, the instrument is the most precise in relative value. If value x characterizes a measurand, the precision of the instrument will be equal to the ratio (dx/x) of global error represented by dx and x. This characterizes the quality of an instrument from the point of view of error. As previously stated, precision is much greater, as indications are closer to the true value (i.e. dx is small). A precise instrument is reliable and accurate at the same time, for example:

For $\mu = 1$; $\sigma = ½$; k = 1 and mean = ($\mu - 4.5\sigma$, $\mu = 4.4\sigma \ldots \mu + 4.5\sigma$)

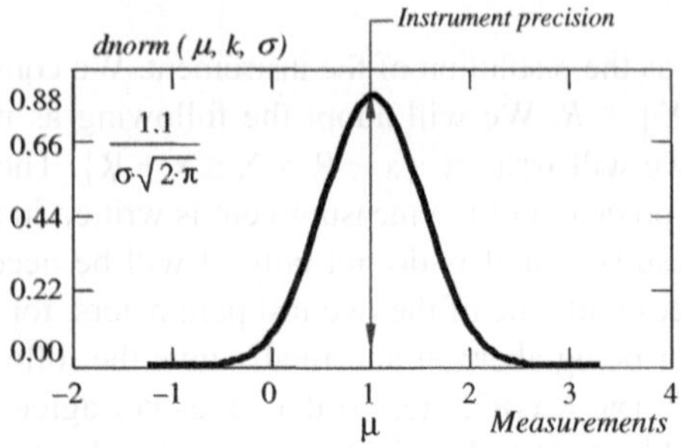

Figure 3.3. *Illustration of the precision curve: true value (arrows) showing the precise measuring instrument*

3.3.1.5.5. Resolution

Resolution is a quantitative expression representing the smallest interval between two elements. Resolution is therefore the smallest indication difference of a display device or of a recording which can be perceived in a significant way. For a digital instrument, we define the resolution as follows:

$$[Class] = \frac{\text{measurement range, (R)}}{\text{number of measurement points, (n)}} \qquad [3.10]$$

The resolution of the digital display of a measuring instrument is a source of uncertainty. Certainly, if ε is the quantification of the instrument, the size value is situated within the interval $[-\varepsilon/2, +\varepsilon/2]$ with a constant probability in the entire interval. It is therefore a matter of a *rectangular* probability law, the errors bring a dispersion of results during repeated measurements. Their statistical treatment allows us to understand the probable value of the measured size and the fixation of uncertainty limits. When the measurement of the same size X has been repeated n times, the results x_1, x_2, ..., x_n, of μ mean value are given. The dispersion of these results is given by the variance or by the standard deviation (σ).

3.3.1.5.6. Reliability

According to VIM, *reliability* is the "ability of a measuring instrument to give related indications during the repeated application of a measurement of size in the same conditions of measurement". Reliability is, in fact, the ability of a measuring instrument to give measurements free from accidental errors (fortuitous where errors are low). It is graphically presented as follows:

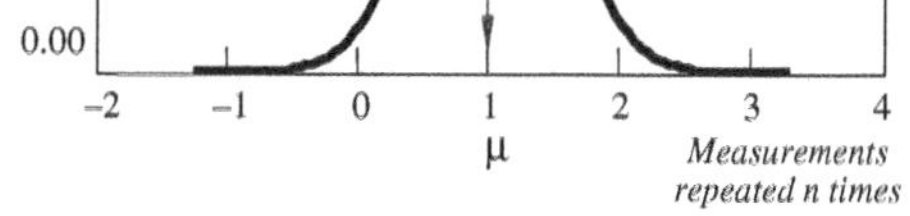

Figure 3.4. *Reliability and true value (the dotted line), showing the instrument of reliability measurement*

For $\mu = 1$; $\sigma = ½$; k = 1; and mean = $\mu - 4.5\sigma$, $\mu = 4.4\sigma$, …, $\mu + 4.5\sigma$.

The standard deviation is considered a reliability error. If we carry out a collection of measurements of size *G*, we obtain a maximum value V_{max} and a minimum value V_{min}. The error limits are:

$$\varepsilon_{\max} = -\left[\frac{V_{\max} - V_{\min}}{2}\right] \text{ and } \varepsilon_{\min} = +\left[\frac{V_{\max} - V_{\min}}{2}\right] \qquad [3.11]$$

3.3.1.5.7. Trueness

Trueness characterizes the ability of an instrument to give indications on the true value of a measured size, untainted by systematic errors. According to the VIM, trueness is defined as: "the ability of an instrument to give indications which are free from systematic error; narrowness of agreement between the mean value of numerous tests and the unknown reference value". The mean result is tainted by trueness errors itself.

The trueness of the result, U = u(t) → Is this a true result in metrology?

We cannot affirm this, due to interfering causes (laws of physics or local experimental approaches). We can predict the effects of these causes, i.e. estimate the value of error component *U*. The variability of the result is inconsistent; hence, we have a consequence on measurement error that is equal to $U - u$. Furthermore, trueness errors are global errors resulting from all causes for each of the measurement results taken in isolation. For multiple measurements, it is the difference between the mean result and the true value. μ is an arithmetic mean of a large number of measurements. V_{true} is the value which is *believed to be true.*

$$D = \mu - V_{true\ value} \qquad [3.12]$$

In order to evaluate trueness, we must use *references.* If the reference value is *traceable* to the international system, according to ISO 3534-1 and ISO 5725-1, the accepted reference value is the value which is *conventionally true.* We speak of *bias* or systematic errors. Bias is described as the "difference between the mean and the accepted reference value". For a reference value $X_{reference}$ with a mean (μ), we express bias as:

$$Bias = \mu - X_{reference} \text{ and } E_{relative\ trueness} = +\left[\frac{\left(\mu - X_{reference}\right)}{X_{reference}}\right] \times 100 \qquad [3.13]$$

Systematic errors are due to the lack of accuracy of the method in operation. *Crude* errors which can be eliminated by the manipulator can also occur.

– e_m measurement error or accuracy;

– bias: systematic error of accuracy;

– inaccuracy or unreliability: random error in metrological repeatability (or reproducibility).

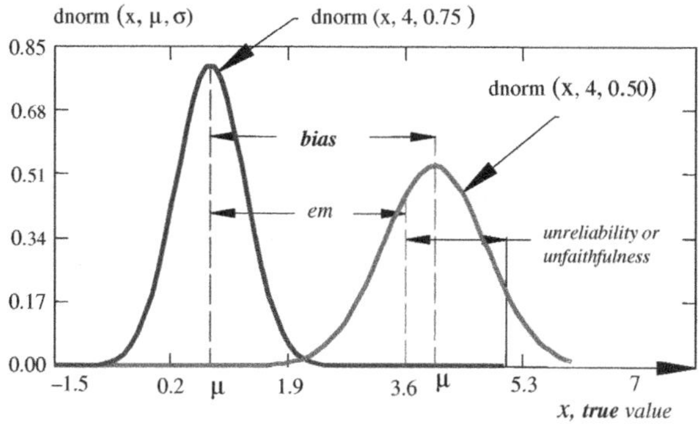

Figure 3.5. *Graphic illustration of bias*

3.3.1.5.8. Accuracy

Accuracy, not precision, expresses the degree of concordance between a *measured value* and the *true value* or expected of the quality of interest. Accuracy (do not use the term precision) is the narrowness of the agreement between the result and the measurement and a true theoretical value responsible for a value of a *recognized reference* illustrated as follows:

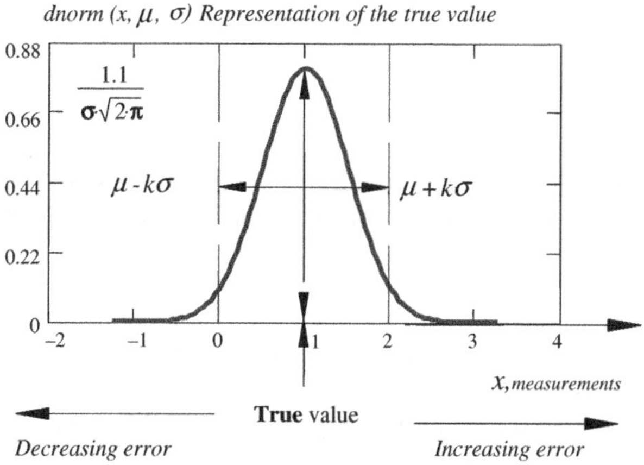

Figure 3.6. *Statistical illustration of error and the "true value"*

For: $\mu = 1$; $\sigma = ½$; k = 2; and mean= $\mu - 4.5\sigma$, $\mu = 4.4\sigma$, …, $\mu + 4.5\sigma$.

DISCUSSION.– Contrary to the "received ideas" of the measured value is the "ultimate truth value". It appears that this fact is not as explicit as it seems. The measured value is never the true value. The following illustration shows how we can imagine this enigmatic true value.

Ø25.030 ← *Value said to be true i.e. forever unknown, e.g. around a diameter of 25.400 mm* → Ø25.035

ATTENTION.– True value does not necessarily indicate the absolute mean. The true value is never known. In fact, it is a collection of values which could reasonably be attributed to the boring value considered in this case.

3.4. Global error and its uncertainty

In conclusion, we will note that errors of different types are brought together. It is often difficult to mark the boundary of their cumulative distribution to get global error from it. It is recommended to group together errors of the same type, but often we infer a global error which takes all errors into account. We will call this the error global algebraic sum of component errors. If we have the following errors:

– δ_a, error due to the instrument: + 0.0005;

– δ_l, measurement error: – 0.0010;

– δ_m, manipulation error: – 0.0060;

– δ_t, error due to the differences of t: + 0.0160 (calculated).

$$\delta_{global} = \sum_{i=1}^{n=4} Errors(+0.0005)-(0.001)-(0.006)+(0.016) = 0.0095mm \qquad [3.14]$$

The correction that must be carried out is opposite in sign [–0.009 5] to the calculated value. The uncertainty will therefore be: ±0.0005 ± 0.001 ± 0.0006 ± 0 = ± 0.005.

The study of uncertainty [MUL 81, TAY 05] also has the objective to determine the capabilities of the means of the measurements. During the setup of a calibration method, it is necessary to proceed with a metrological qualification of the method. This operation is based on technical tests and on the objective analysis of causes of uncertainty. Uncertainty which is linked to an instrument to be calibrated is determined based on the instrument's own characteristics, notably including reliability and measurement error. The capability of measurement instruments is an important notion because it provides information on the degree of concordance, which connects the performance of an instrument to the tolerance to be verified.

Capability is the adequacy between the repective tolerance interval and the global uncertainty of measurement. Capability methods have been developed by the Automotive Industry Action Group (AIAG) in the USA. They are applicable to other sectors of industry and do not contradict GPS norms. We know from experience that a choice of over-performing measurement techniques inevitably proves prohibitive and leads to over-quality (Chapters 1, 2 and 3 volume 3). Some dimensions which come from the definition design with harsh tolerances are difficult to put into use. They are rejected during testing. The choice of instrument is subject to the tolerance to be verified.

$$C_p = \left\{ \frac{T_s - T_i}{6 \times \sigma} \right\} \qquad [3.15]$$

where

T_S is the superior tolerance

T_I is the inferior tolerance

σ is the standard deviation of the series of parts

μ is the mean of the series

The coefficient of capability C_{pk} will be:

$$C_{pk} = \left\{ \frac{TS - \mu}{3 \cdot \sigma} \right\} \text{ and } \left\{ \frac{TI - \mu}{3 \cdot \sigma} \right\} \qquad [3.16]$$

where

TI is the tolerance interval

UG is global uncertainty

The capability indicator C_{mm} of the measurement means 6σ is written as:

$$C_{pk} = \left\{ \frac{IT}{6 \times UG} \right\} \text{ where UG is global uncertainty} \qquad [3.17]$$

3.5. Definitions of simplified equations of measurement uncertainty

Uncertainty measurement (*U*) is a parameter associated with the measurement result which characterizes the dispersion of values which are reasonably attributed to the object of measured value. The uncertainty of this measure constitutes the result of the combination of component effects and the sources of uncertainty, commonly called scales of influence. Hence, the ISO/IEC 17025 norm mentions uncertainty estimation. According to the GUM, global uncertainty includes all influences [NIS 94, NIS 97]. Composed uncertainty results from the combination of calculation of uncertainties according to the propagation law as follows:

– Standard uncertainty is generally expressed under the form of a standard deviation (S – standard deviation).

– Standard deviation is the dispersion of the results of n measurements of the same size around the arithmetic mean x of n results. x_i being the result of the ith measurement.

– The coefficient of variation S_R indicates the standard deviation divided by the mean (S/μ). It is standard deviation as a relative value (%) and not the absolute value in the unit of measurement.

Repeatability is the narrowness of agreement (expressed in the form of the standard deviation) between the results of successive measurements of an identical object responsible for an equally measured scale in identical measurement conditions.

The measurements are taken in repeatability conditions, i.e. same procedure, same operator, same apparatus (instrument) used in the same conditions, in the same place, at the same time, and with the same quantitative expressions of the result.

Reproducibility is the narrowness of agreement (generally expressed in the form of the standard deviation) between the results of successive measurements of the same object responsible for the same size, measured in different measurement conditions (more specifically, principal, method, operator, instrument, material (standard) of reference, place, conditions, time (date), quantitative expression of the result).

Expanded uncertainty describes an interval around the result of measurement from which we can expect comprehension of an elevated proportion of the distribution of values, reasonably attributed to a measured size. It is a multiple of the standard deviation S or of global uncertainty U by a factor of expansion, k. $k = 2$ statistically signifies that the value which is "reasonably attributable" to the measured object has a probability or a confidence threshold of around 95% in the interval, more or less than two times S is responsible for two times U around the measured value. With a factor $k = 3$, the confidence level is at around 99.7% on a normal distribution. The expansion factor (coverage) k is a multiplicative factor of the *uncertainty of the composed sort* to obtain *expanded uncertainty*.

In Figure 3.7, *mp* is the production margin; U(x) the measurement uncertainty; *env.* is a deviation due to the environment; *mmm* is a metrology maneuver margin; and *gmm* is a global maneuver margin.

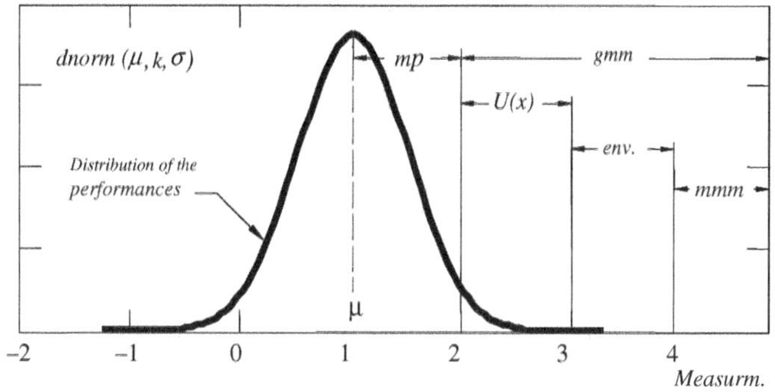

Figure 3.7. *Schematic illustration of the specifications of measurements*

By combining them, various uncertainties lead to what is called *overall uncertainty*. It is a *combined global uncertainty* (CGU written as $u^2_c(y)$). A CGU is a global polyfactorial function which is, in fact, a result with all the sizes of input X_i. We note, however, that in numerous cases, the measurand does not directly come from, but is rather determined based on, n polyfactorial input values of function Y:

$$Y = f\left\{X_1, X_2, X_4, \ldots, X_n\right\} \quad [3.18]$$

This relation does not only translate a physical law, but also the measurement procedure. The translation of all the values which significantly contribute to uncertainty results in the form of $Y = kx$ (unit). The simplified form of the expression from $u_c^2(y)$ in the practical dimensional metrology case is brought back to the simple forms shown in the following equation. By A, we mean the sum of X_i quantities multiplied by constant factors a_i giving:

$$Y = a_1X_1 + a_2X_2 + a_3X_3 + \ldots\ldots + a_nX_n \quad [3.19]$$

The results of various measurements are written as:

$$y = a_1x_1 + a_2x_2 + a_3x_3 + \ldots\ldots + a_nx_n \quad [3.20]$$

Each of these measurements is affected by uncertainty. Combined uncertainties are expressed as:

$$u_c^2(y)=a_1^2\cdot u^2(x_1)+a_2^2\cdot u^2(x_2)+a_3^2\cdot u^2(x_3)+......+a_n^2\cdot u^2(x_n) \quad [3.21]$$

The measurement equation will be written for the product of X_i quantities raised to power a, b, ..., q, multiplied by the constant A:

$$Y = Cste(A)\times X_1^a\cdot X_2^b\cdot X_3^c\cdot.....\cdot X_n^q \quad [3.22]$$

The CGU will be composed and will take the form:

$$y = Cste(A)\times x_1^a\cdot x_2^b\cdot x_3^c\cdot.....\cdot x_n^q \quad [3.23]$$

Composed uncertainty is expressed as:

$$u_{cr}^2(y)=a^2\cdot u_{cr}^2(x_1)+b^2\cdot u_{cr}^2(x_2)+c^2\cdot u_{cr}^2(x_3)+...+q^2\cdot u_{cr}^2(x_n) \quad [3.24]$$

In this case, $u_r(x_i)$ expresses uncertainty relative to x_i and is defined by:

$$u_r(x_1)=\frac{u(x_i)}{|x_i|} \quad [3.25]$$

$|x_i|$ is the absolute value of $x_i \neq 0$. $u_{c,r}(y)$ is relative combined uncertainty, written:

$$u_{cr}(y)=\frac{u_c(y)}{|y|} \text{ with } |y| \text{ which is an absolute value of } y\neq 0 \quad [3.26]$$

3.5.1. *Expansion factor k and range of relative uncertainty*

If the distribution probability characterized by the result of measurements y and $u_c(y)$, its standard deviation and its uncertainty type combined are approximated by the expression of a Gaussian distribution, and if $u_c(y)$ is estimated, then the interval $\{y - u_c(y)\}$ to $\{y + u_c(y)\}$ contains ~68% of the distribution of the true value. We will notice in the 68% confidence interval that Y is greater than or equal to $y - u_c(y)$, and lees than or equal to $y + u_c(y)$, often noted using:

$$Y = y\pm u_c(y) \quad [3.27]$$

The combined expanded uncertainty range u_c is used to express the uncertainty of various measurements and their regularity, as in material and structure reliability [GRO 94]. In this case, we obtain the uncertainty range using U which is obtained by multiplying $u_c(y)$ by a coefficient called the expansion factor k (*coverage factor*). The expression of U is written:

$$U = k \cdot u_c(y) \tag{3.28}$$

With $Y \geq \{y - U\}$ and $\{y + U\}$, we postulate the expression

$$Y = y \pm U \tag{3.29}$$

Expansion factor k is generally chosen according to the desired confidence interval, in agreement with the confidence interval defined by:

$$U = k \cdot u_c \tag{3.30}$$

Expansion factor k is of rank 1, 2, and 3 when the distribution law is normal and u_c are the reliable estimators of the standard deviation of y. For example:

– $U = 1 \times u_c$ ($k = 1$) will be defined in a confidence interval of 68%.

– $U = 2 \times u_c$ ($k = 2$) will be defined in a confidence interval of 95%.

– $U = 3 \times u_c$ ($k = 3$) will be defined in a confidence interval of >99%.

By analogy with the range of relative uncertainty u_r and the standard uncertaintypreviously defined by the relation $u_c^2(y)$, the uncertainty range relative to y, for $y \neq 0$, measurements are calculated by:

$$U_r = \frac{U}{|y|} \tag{3.31}$$

The method of measurement, the instruments used, as well as the abilities of the experimenter each contribute to the size of σ. This choice aims to avoid unnecessary mistakes. When these quantities are used to express components of uncertainty, we use the following notations:

– variance (X) = VA(X) = u^2(X);

– uncertainty of type u(x) = $\sqrt{u^2(x)}$;

– composed uncertainty = u_c(y);

– expanded uncertainty U = $\kappa.u_c$(y), (κ, expansion coefficient);

– uncertainty range Y = y ± U.

DISCUSSION.– It has been succinctly observed that the uncertainty range depends directly on combined global uncertainty. The simplified form of the expression which comes from $u_c^2(y)$ in the practical metrology case is brought back to its simple form.

3.5.2. *Determination of type A and B uncertainties according to GUM*

The ISO norm of the VIM describes measurement uncertainty as a “parameter associated with the result of a measurement which characterizes the dispersion of values which could reasonably be attributed to the measurand”. Various guides including the GUM [GUI 00, NIS 94] clearly describe how this is possible:

– evaluating the contribution of each source of uncertainty separately;

– combining different contributions;

– declaring the uncertainty of a measurement’s result.

A definitional uncertainty imposes an lower limit on all measurement uncertainties. The interval can be represented by one of these values, called the *measured value*. In the GUM, *definitional uncertainty* is thought to be negligible compared to measurement uncertainty considered. The *measurand* can therefore be represented by a value using *unique essence*. The objective of measuring would be to establish the *probabilities* that the measured values are compatible with the definition of the measurand. The GUM defines two methods of uncertainty evaluation: *type A* uncertainty which uses statistical means with repeated measurements and the calculation of standard deviations and *type B* uncertainty which uses other earlier measurement standard data, repeatability, reproducibility, and *intercomparisons*, published constants. The data are in the form of standard deviation or of intervals, or from *a priori* law.

Composed uncertainty is calculated by combining uncertainties according to the law of propagation of the calculation function. Compound uncertainty $U(y)$ (excluding correlations) is written as:

$$U_{y(compound)}^2 = \sum_{i=1}^{n} \left(\frac{\partial f}{\partial x_i} \right)^2 \cdot U_{xi}^2 \qquad [3.32]$$

The partial derivatives $\partial f/\partial x_i$ are called *sensitivity coefficients* expressed by dispersion. Compound uncertainty is the expression of the following standard deviation:

$$U_{y(compound)} = \sqrt{\sum_{i=1}^{n} \left(\frac{\partial f}{\partial x_i}\right)^2 \cdot U_{xi}^2} \qquad [3.33]$$

The properties of the compound uncertainty functions [DIX 51, GUI 00] are:

i) For $Y = A + B$, $U_c(y) = \sqrt{U_A^2 + U_b^2}$.

ii) For $Y = A - B$, $U_c(y) = \sqrt{U_A^2 + U_b^2}$.

iii) For $Y = A \times B$, $U_c(y) = \sqrt{B^2 \cdot U_A^2 + A^2 \cdot U_b^2}$.

iv) For $Y = A / B$,: $U_c(y) = \sqrt{(U_A^2 / B^2) + A^2 \cdot (U_b^2 / B^4)}$.

v) For $Y = k \times A$, $U_c(y) = k \times U_A$.

The experiment has shown that uncertainties which results due to the analytical method are often too inaccurate. We use repeated measurements for each size of influence by means of equations, which are sometimes complex. The parameter may be a standard deviation (or a multiple of this) or the half-width of an interval of a determined level of confidence. Before any of this, it is useful to make a linear combination of the uncertainties, i.e. applying the law of propagation while considering that all the uncertainties are correlated. The work consists of determining each quantity x_i as well as the uncertainty of type $u(x_i)$ which is associated with it.

It is the convention to use the absolute values of U for such coefficients of dimensional sensitivities. On the other hand, we use values which are relative in the expressions, integrating the sensitivity coefficients (C generally is unitary (= 1), i.e. adimensional). The composed uncertainty evaluation is done with the help of rules of the propagation of errors or by default according to the appropriate formulae (*a priori* rectangle law). The GUM approach is succinctly resumed as follows:

The propagation law of uncertainty allows us to calculate the composed variance $u_c^2(y)$. From this follows the composed standard deviation $U_c(y)$ and then the expanded uncertainty (U), which is obtained by multiplying the composed standard deviation by an expansion factor k. The expansion factor value is linked to the desired probability. The uncertainties are evaluated according to their different components. In all evaluation processes of measurement uncertainty, we are led to estimate the uncertainties of type $u(x_i)$ or the corresponding variance $u_c^2(x_i)$ to each of the components which would intervene in the evaluation of composed uncertainty $U_c(y)$.

– Can't the measured sample represent the finite measurand?

– Is the knowledge of the environmental effects on the measurement procedure correct?

– Is the resolution of the instrument correct?

– Have we assigned values and reference materials to the standards?

– Have we properly estimated the hypotheses retained in the measurement procedures?

– Are the conditions of the repeated observations of the measurand identical?

No.	Analytical approach of (GUM) simplified in eight steps
1	Formulate the result y contingent on the input values $(x_1, x_2, ..., x_n)$ $y = f(x_1, x_2, ..., x_n)$
2	Determine the data/input values (measurements, specification data)
3	Determine the uncertainty of each piece of data/input value of type A and/or B
4	Identify the covariances, i.e. the correlations between the effects of different types of uncertainty on the data/values, by default. We generally disregard correlations, which can falsify uncertainty calculations
5	Calculate the result $f(x_1, x_2, ..., x_n)$ of the measurement with the input values $(x_1, x_2, ..., x_n)$
6	Calculate the composed uncertainty with the data from point 3, excluding correlations. To simplify, we calculate them according to laws of propagation of uncertainty contingent on the mathematical formula of the result
7	Calculate the expanded composed uncertainty with $k = 2$, (e.g. 2 times the composed uncertainty)
8	Deliver the result of measurement y with expanded, composed uncertainty. Signify $k = 2$, for a "confidence level" of ~95% for a Gaussian distribution, e.g. $k.U_c(y)$

Table 3.1. *Summary of the GUM approach in eight distinct steps*

It is beneficial to consider each case as if it was unique to the laboratory where the measurement experiments take place. It is sensible of us to have proposed a few of the different components of uncertainty in the manual [GUI 00] ISO/CEI/OIML/BIMP. During the calculation of the uncertainty measurement, a metrologist's problem is to identify the sum of the x_i components which have an impact on the measurement result and to quantify their uncertainty type.

To evaluate the numerical value of this final element, the BIPM proposes two frequently used methods: the *type A* method and the *type B* method. *Type A* methods are based on the variability of the measurement result, hence the statistical tool used to analyze them (*repeatability conditions* and *reproducibility conditions*, explained earlier). It is a global statistical approach of n measurements.

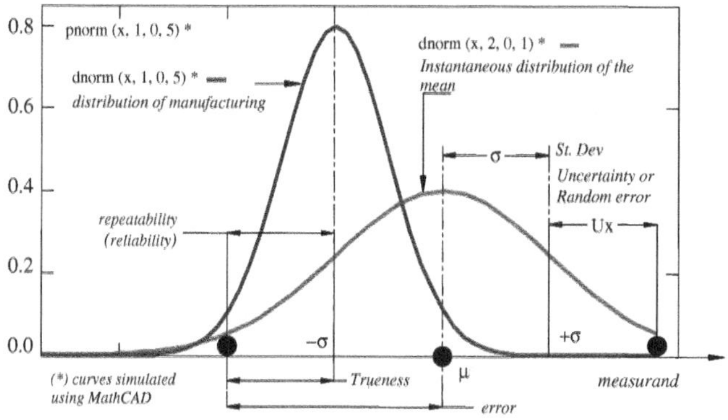

Figure 3.8. *Graph which is useful to the field of study using methods A and B*

The mean $\mu = E(x)$ represents the result of the measurement, and the standard deviation constitutes the uncertainty u_x of one series of n measurements repeated from the (x_1, x_2, x_3, ..., x_i, ..., x_n) input values. Ideally, n tends toward infinity (∞). So, we practically have a sample to statistically analyze using a representative distribution law.

Type B methods do not use the statistical tool, which is indispensable for the causes which do not bring about the variation of the polyfactorial result. We will draw up a report of the errors. This report should include systematic errors such as the parallax error, the tuning to zero of the instrument, etc., or random errors, such as measurement errors or errors that come from the instrument itself, etc. We use physics or associated hypotheses of the distribution law, named *a priori* law (rectangle, for example). In fact, this method covers everything which is not statistical (specification, calibration certificates, influence factor, etc.).

3.5.2.1. *Succinct description of the type A uncertainty method*

The evaluation of type A uncertainty applies to bias error. The only condition is that the calculation of the uncertainty component will be based on a statistical analysis of the data. The following distinction regarding random error and partiality should be borne in mind:

– Random errors cannot be corrected.

– Bias can, theoretically be corrected or eliminated by the result.

In material and structural reliability, one of the most important indicators of random error is *time*. Uncertainty makes us realize all the differences which are dependent on *instruments*, *operators*, *geometry*, *etc.* The causes of the differences within a laboratory are:

1) differences between the measurement instruments (derived units), e.g. the instruments cannot be calibrated directly to a reference base (silicon);

2) differences between operators for optical measurements which are not automatized and are heavily dependent on operator observation;

3) differences between geometrical configurations of instrumentation.

With respect to time, changes are a primary source of random errors. One of the most important indicators of random error is therefore time. Three levels of error which are dependent on time are as follows:

– Level I or short-term errors (repeatability, imprecision, etc.);

– Level II or common errors (reproducibility);

– Level III or long-term errors (stability).

The best approach for collecting information on time-dependent sources consists of intercalculating the workload with the measurements. As these are time-dependent, the sources of error are evaluated and quantified on the verification database of standard measurements. It is a long-term *bias* and *variability* control system. In these same experimental conditions, ***repetitions*** give rise to mathematical dispersions on numerical values of measurements repeated (***n***) times. We assume that the measurement procedure uses a good *resolution*. The arithmetic mean μ calculated based on individual x_i values allows a good estimation of the mean of the population in consideration to be given, when we consider ***n*** values independent v_{ij}.

The repetition of the *n* measurements is approached by the arithmetic mean (μ). This allows the comprehension of all of the x_i values, as being a random variable constituting a random variable (RV). We will calculate the variance on the arithmetic mean by applying the law of propagation of the uncertainties [NIS 94].

$$U_c^2(\mu)=\left(\frac{1}{n}\right)^2\times u^2(x_1)+\left(\frac{1}{n}\right)^2\times u^2(x_2)+\left(\frac{1}{n}\right)^2\times u^2(x_3)+\ldots+ \\ 2\times\sum_i\sum_j\left(\frac{1}{n}\right)\cdot\left(\frac{1}{n}\right)\cdot\left\{u\left(x_i\cdot x_j\right)\right\} \quad [3.34]$$

Let us now express the correlation coefficient between the two values x_i, and x_j using *R*:

$$u\left(x_i,x_j\right)=R\left\{u\left(x_i\right),u\left(x_j\right)\right\} \quad [3.35]$$

We notice that in the case of non-correlation, i.e. independent probabilities, the Monte Carlo (MC) simulation [GUI 00] would be the most appropriate to our study. If $u(x_i)$, $= \sigma^2$ is the variance of the population of a series of experiments, after simplification, we will have:

$$u^2(\bar{x}) = \left\{ \left(\frac{1}{\kappa} \right)^2 \times \kappa\sigma^2 + \kappa(\kappa - 1) \times \left(\frac{1}{\kappa^2} \right) \times \sigma^2 R \right\} = \left\{ \frac{\sigma^2}{\kappa} + \left(\frac{\kappa - 1}{\kappa} \right) \times \sigma^2 R \right\} \quad [3.36]$$

We notice that when the observations on the measurements are independent, $R = 0$; hence [3.36] will be written as:

$$u^2(\bar{x}) = \left\{ \sigma^2 \frac{1}{\kappa} \right\} \quad [3.37]$$

When the observations on the measurements are totally correlated, i.e. $R = 1$, then expression [3.37] will take the following form:

$$u^2(\bar{x}) = \sigma^2 \quad [3.38]$$

The estimator is therefore translated using the expression following the mean:

$$x_j = \frac{1}{\kappa} \cdot \sum_{i=1}^{\kappa} x_{ij} \quad [3.39]$$

So, the experimental estimator of the variance is expressed by:

$$\sigma_{xj} = \sqrt{\left(\frac{1}{\kappa - 1} \right) \times \sum_{i=1}^{\kappa} \left(x_{ij} - \bar{x}_{ij} \right)^2} \quad [3.40]$$

The expression of the standard deviation will be represented by the following relation:

$$u(x_i) = \sigma(\bar{x}) = \sqrt{\left(\frac{1}{\kappa - 1} \right) \cdot \sum_{i=1}^{\kappa} \left(x_{i,\kappa} - \bar{x}_i \right)^2} \quad [3.41]$$

If, in a case of laboratory figures, several experiments $\{k_1, k_2, \ldots, k_i\}$ had been conducted and the corresponding estimators: $\{S_1^2, S_1^2, \ldots, S_i^2\}$ had been calculated, the expression of the variance of the total population could be obtained by combining these different estimators to obtain the following relation:

$$\sigma^2 = \left\{ \frac{(\kappa_1 - 1)\times\sigma_1^2 + (\kappa_2 - 1)\times\sigma_2^2 + ... + (\kappa_i - 1)\times\sigma_i^2}{(\kappa_1 - 1) + (\kappa_2 - 1) + ... + (\kappa_i - 1)} \right\} \quad [3.42]$$

By introducing the number of df $\lambda_i = df$ such that $\{\lambda_I = k_i - 1\}$ the estimator is written:

$$\sigma^2 = \left\{ \frac{\lambda_1\sigma_1^2 + \lambda_2\sigma_2^2 + ... + \lambda_i\sigma_i^2}{\lambda_1 + \lambda_2 + ... + \lambda_i} \right\} \quad [3.43]$$

This method thus allows the calculation of the *repeatability* component $u^2(\mu)$:

$$u(\overline{x})^2 = \frac{(\sigma^2)}{\kappa} \quad [3.44]$$

If in a given measurement procedure, the operator carried out a single measurement $u(\mu)^2$:

$$u^2(\overline{x}) = \sigma^2 \quad [3.45]$$

Relations [3.44] and [3.45] indicate that we must estimate the repeatability of the measurement procedure using several preliminary tests before proceeding. We notice, thanks to the above, that the *type A* method puts pressure on the statistical approach due to the multiplicity of the input values.

3.5.2.1.1. Sensitivity coefficients for the type A components of uncertainty

This section defines the sensibility coefficients which are appropriate to estimated components based on type repeated measurements. Common errors can be the main source of uncertainty. With an instrumentation which is extremely precise in the short term, changes in time, often caused by small environmental effects, are often the main source of uncertainty in the measurement process. Two levels of error dependent on time are probably sufficient to describe the majority of measurement processes. Three levels may be necessary for new measuring or procedure processes whose characteristics are not well understood. Measurements repeated on the test point do not generally cover a time lapse which is sufficient to capture daily variations in the measurement process. The difference in these measurements is cited as the estimation of the uncertainty of no piece of data being available for evaluation. For short-term measurements, J, this difference has $\nu = (J - 1)$ df.

3.5.2.1.2. Overlapping conception for the estimation of type A uncertainties

A less efficient method for estimation, dependent on uncertainty time sources, is an experiment which has already been thought of. Measurements can be carried out specifically to estimate two or three levels of error. Several ways of doing this exist, but the simplest method is an overlapping conception where the *j* of short-term measurements is replicated on *k* days and the whole operation is then replicated on complying with (*L*) (month, etc.). The analysis of these data leads to:

σ_1 = difference ($J - 1$) df, for short-term errors;

σ_2 = difference ($K - 1$) df, for common errors;

σ_3 = standard deviation ($L - 1$) df, for very long-term errors.

Coefficients of sensitivity – case study for the propagation of error: if the uncertainty of the final value is calculated based on the propagation of errors, the sensitivity coefficients constitute the multipliers of the individual terms of variance. Formulae are given for certain functions with one or several variables. The standard deviation of signaled values which are the functions of a single variable are reproduced in Table 3.2 [KU 66]. The final value, *Y*, is a function of the mean of *n* measurements on a single variable.

Observations	Function *Y* of mean μ of n measurements, $Y = f(\mu)$	Standard deviation of *Y*, σ_x = standard deviation of *X*
Y is the function representative of the physical process (experimental)	$Y = \mu = \bar{X}$	$\sigma_x/\sqrt{N}$
	$Y = \left(\frac{\bar{X}}{1+\bar{X}}\right)$	$\frac{\sigma_x}{\sqrt{N}\left(1+\bar{X}\right)^2}$
	$Y = \left(\bar{X}\right)^2$	$\left(\frac{2\bar{X}}{\sqrt{N}}\right)\sigma_x$
	$Y = \sqrt{\bar{X}}$	$\sigma_x/2\sqrt{N\times\bar{X}}$
	$Y = \ln\left(\bar{X}\right)$	$\sigma_x/\left(\bar{X}\sqrt{N}\right)$
If (*n*) is small, the approximation could be compromised	$Y = Exp^{\bar{X}}$	$\frac{Exp^{\bar{X}}}{\sqrt{N}}\sigma_x$
Not directly derived from formulae	$Y = \left(100/\bar{X}\right)\sigma_x$	$Y/\sqrt{2(N-1)}$ We estimate that the original data follow a normal distribution

Source: Harry KU [KU 66].

Table 3.2. *Calculation formulae (standard deviation)*

Function to two variables: The final value Y is a function of the means of (n) measurements out of two variables: Y = f(X, Z)

Function: Y of $\overline{X},\overline{Z}$ $\overline{X}$ and $\overline{Z}$ are the means of n measurements	Standard deviation of Y, σ_x = standard deviation of X σ_z = standard deviation of Z	$\sigma_{xz}^2 = COV(X, Z)$ The term Cov from Covariance is included exclusively for a weak estimation
$Y = A\overline{X} + B\overline{Z}$	$\frac{1}{\sqrt{N}}\sqrt{A^2\sigma_x^2+B^2\sigma_z^2+2AB\sigma_{xz}^2}$	
$Y = \frac{\overline{X}}{\overline{Z}}$	$\frac{1}{\sqrt{N}}\left(\frac{\overline{X}}{\overline{Z}}\right)\sqrt{\frac{\sigma_x^2}{\overline{X}^2}+\frac{\sigma_z^2}{\overline{Z}^2}-2\frac{\sigma_{xz}^2}{\overline{X}.\overline{Z}}}$	
$Y = \frac{\overline{X}}{\overline{X}+\overline{Z}}$	$\frac{1}{\sqrt{N}}\times\left(\frac{Y}{\overline{X}}\right)^2\sqrt{\overline{X}^2\sigma_x^2+\overline{Z}^2\sigma_z^2-2\overline{X}\overline{Z}\sigma_{xz}^2}$	
$Y = \overline{X}\cdot\overline{Z}$	$\frac{\overline{X}\overline{Z}}{\sqrt{N}}\sqrt{\frac{\sigma_x^2}{\overline{X}^2}+\frac{\sigma_z^2}{\overline{Z}^2}+2\frac{\sigma_{xz}^2}{\overline{X}.\overline{Z}}}$	
$Y = c\left(\overline{X}\right)^a\cdot\left(\overline{Z}\right)^b$	This is an approximation. The exact result is obtained with the help of the exact formula [GOO 60] of Goodman for the standard deviation of a derived probability	

Source: Harry KU [KU 66].

Table 3.3. *Formulae of calculations*

Function with several variables: The propagation of error for several variables can be simplified if the function Y is a simple multiplicative function of secondary variables or if uncertainty is evaluated in percentage. For example, for the three variables X, Z, W, the function Y = (X · Z · W) has a standard deviation in absolute units of:

$$\sigma_Y = \sqrt{(ZW)^2\,\sigma_x^2 + (XW)^2\,\sigma_z^2 + (XZ)^2\,\sigma_w^2} = \sqrt{\frac{\sigma_x^2}{X^2}+\frac{\sigma_z^2}{Z^2}+\frac{\sigma_w^2}{W^2}} \quad [3.46]$$

In absolute terms (%), the difference can be described as:

$$\frac{\sigma_Y}{Y} = \sqrt{\frac{\sigma_x^2}{X^2}+\frac{\sigma_z^2}{Z^2}+\frac{\sigma_w^2}{W^2}} \quad [3.47]$$

If all the covariances are negligible. These formulae are easily spread out with more than three variables. The coefficients of sensibility and of standard deviations are combined by the sum of the square root to obtain a "standard uncertainty". Given the R components, uncertainty is normalized:

$$u = \sqrt{\sum_{i=1}^{R} \psi_i^2 \times s_i^2} \qquad [3.48]$$

Expanded uncertainty ensures a increased confidence level. If the object of the declaration of uncertainty is to supply a coverage with an increased level of confidence, an expanded uncertainty is calculated as an earlier equation: $U = \kappa \times u$. κ is chosen to be the critical value ($\alpha/2$) of Student's t-table with v df. For high df, $\kappa = 2$ approaches a coverage of 95%. The expanded uncertainty defined earlier is assumed to supply a high level of coverage for the true unknown value of the measurement of interest which provokes it; this means that for any measure, Y, $\mathrm{Y} - \mathrm{U} \leq \text{true value} \leq \mathrm{Y} + \mathrm{U}$.

To express measurement uncertainties, the ISO guide [ISO 97b] assumes that all bias is corrected for interlaboratory tests and that uncertainty applies to the corrected results [ISO 97]. For measurements at the level of industry, this approach presents several disadvantages. Corrections can be costly to carry out if they require modifications to existing computer programs and corrections can be fastidious and subject to caution. In cases such as these, the best way of proceeding is to signal the measurements taken and to adjust uncertainty to take "bias" into account. The main issue is knowing how to adjust uncertainty [PHI 97]:

1) The final uncertainty must be greater than or equal to the considered uncertainty if bias is to be corrected.

2) The final uncertainty must reduce uncertainty itself, due to the correction of bias.

3) The coverage level obtained from the final declaration of uncertainty should be at least the level obtained for the corrected bias case.

4) The method must be transferable for uncertainty and bias to use uncertainty components in another declaration of uncertainty.

3.5.2.2. *Type B uncertainty methods*

We use these methods to quantify the uncertainties of different components which intervene in the measurement process. Here there are uncertainties relating to *calibration* corrections and even those linked to the *environment*. *Type B* methods [NIS 94] are often used when we want to free ourselves from statistical approaches. It is for this reason that many lab assistants, not very inclined toward the mathematical sciences, prefer this method. It depends on the experience acquired by the operators since they also rely on tests and the knowledge of physical phenomena. To study these methods, we focus on a concrete *experimental* case, on the measurement process. For each x_j variable involved in the measurement process, we will formalize the uncertainties of corresponding types, averaging the characteristics previously mentioned. For this to happen, we must know the distribution law of the measured variables before doing anything, as well as the range of these values.

In fact, type *B* methods are not as experimental as certain pieces of literature imply. They perform the task of abstraction of statistical methods in themselves correctly, but they make the most of the theoretical results of these methods. In this way, if the manufacturer of the instrument gave the uncertainty type, it would be appropriate to use it. For example, if the tolerance of a graduated rule of 30 mm is 0.025 mm, the uncertainty will be $U = U_r$ for a *type B* error:

Of $U_r = \frac{U}{|y|}$ for an initially uniform distribution $U_r = \frac{\Delta c}{\sqrt{3}} = \frac{0.025}{\sqrt{3}} = 0.014$.

If the error is approached by a Gaussian distribution, a situation which is quite common (cumulative errors), we will apply a normal distribution law which calculates uncertainty using *U*. For Δc = 0.025, see Table 3.6 for a Gaussian distribution law:

$$U = \frac{\Delta c}{3} = \frac{0.025}{3} = 8.333 \times 10^{-3}$$

The evaluation of *type B* uncertainty applies to random errors. The calculation of the standard deviation depends on the number of *repetitions* on the testing point and the range of environmental and operational conditions during which the repetitions were carried out. Other sources of error will be considered as standardization uncertainties for the measurement reference standards, which influence the final result. If the value of the test element cannot be measured directly, but must be calculated starting with measurements on secondary quantities, the equation to combine different quantities must be defined. The steps to follow are:

1) declaring a value implies quantity measurements;

2) calculating a standard deviation for random sources of error to

i) reproduce the results of a tested element;

ii) take measurements based on a verification norm;

iii) take measurements in line with a certain level (e.g. level II or level III).

If, for example, the temporal components are estimated based on a conception (*design*) overlapped by *level II*, the final value for a test element is a mean (μ) on:

i) short-term repetitions of n;

ii) a daily number (M = 1 is permitted) of days;

iii) measurements of test elements, the standard deviation (σ) of the final value is:

$$\sigma_{reported\_value} = \sqrt{\left(\frac{1}{M}\right)\sigma_{days}^2 + \left(\frac{J-N}{MN}\right)\sigma_1^2} \quad [3.49]$$

where:

M is the daily number;

N is a "centered" number in the short term;

J is a number corresponding to the days;

σ_j are the standard deviations (days).

If the *df* are necessary for the uncertainty of the final value, the above formula cannot be used directly. It is necessary to rewrite the expression of the standard deviation σ_1 and σ_2 explained as follows:

$$\sigma_{reported\_value} = \sqrt{\left(\frac{1}{M}\right)\sigma_2^2 + \left(\frac{J-N}{MNJ}\right)\sigma_1^2} \quad [3.50]$$

So the sensitivity coefficients will be:

$$I = \sqrt{\left(\frac{J-N}{PMNJ}\right)} \; and \; \psi_2 = \sqrt{\frac{1}{M}} \quad [3.51]$$

The specific sensitivity coefficients are indicated in Table 3.4 for the selections of (N) and (M).

Sensitivity coefficients for two uncertainty components			
Number in the short term, N	Daily number, M	Short-term sensitivity coefficient, ψ_1	Daily sensitivity coefficient, ψ_2
1	1	$\sqrt{\left(\frac{J-1}{J}\right)}$	1
N	1	$\sqrt{\left(\frac{J-N}{NJ}\right)}$	1
N	M	$\sqrt{\left(\frac{J-N}{MNJ}\right)}$	$\sqrt{\frac{1}{M}}$

Source: *Engineering Statistics Handbook* (USA).

Table 3.4. *Specific sensitivity coefficients for the selections of* (N) *and* (M)

3.5.2.2.1. Level III: sensitivity coefficients for the measurements of a design of level III

If the temporal components are estimated based on an overlapping conception of level III and the final value is an average of short-term repetitions of n, then M implies days and P executes measurements based on test elements.

The standard deviation of the final value is written as follows:

$$\sigma_{reported\_value} = \sqrt{\left(\frac{1}{P}\right)\sigma_{active}^{2} + \left(\frac{1}{PM}\right)\sigma_{days}^{2}\left(\frac{1}{PMN}\right)\sigma_{1}^{2}} \qquad [3.52]$$

3.5.2.2.2. Analysis of the variability of an overlapping conception (design)

The purpose is to show the effect of different levels of effect dependent on the time of the variability of the measurement procedure with standard deviations for each level of overlapping conception at level III.

– Level I corresponds to repeatability (short-term precision).

– Level II corresponds to reproducibility (from day-to-day = daily).

– Level III corresponds to stability (r & r = run-to-run).

The possible scenarios for a level II design (short-term and daily repetitions) to illustrate the concepts is graphically represented in Figure 3.9. For example, the representation of measurement two presents short-term variability even for six days when process one has a large amount of variability between days and process two has variability between negligible days.

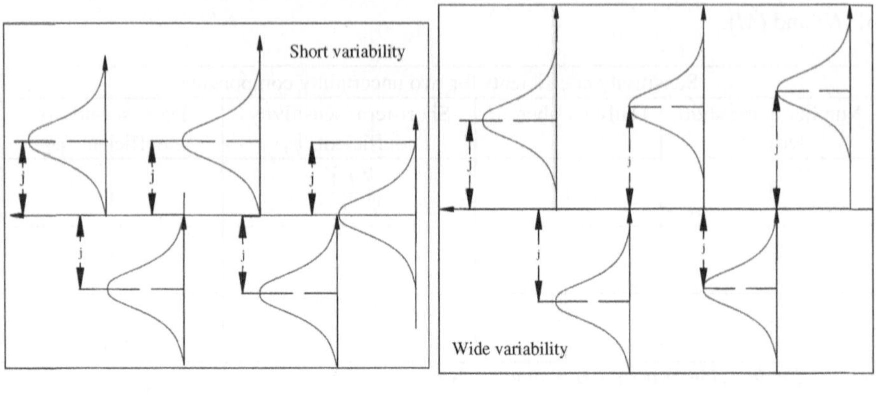

Figure 3.9. *The graph of possible scenarios for a design of level II*

a) Distributions of short-term measurements for more than six days, where the trait distances illustrate variability between days: an easy way of beginning with a table of level II (j) columns and κ lines for repeatability and productability measurements and of proceeding by calculating:

1) the mean for each line and inserting it in the [J + 1]column;

2) the difference of level I (repeatability) for each line and inserting it in the [J + 2] column;

3) the large mean and difference of level II of the data in the[J + 1] column;

4) repeating the table for each of the (l) implemented.

b) Calculation of the difference of level III based on the means of large *L*

Level I: 1κ repeatability, differences can be calculated based on the data. The measurements of overlapping conception are noted using equations corresponding to the tabular analysis. At level I, the repeatability differences, σ_{1lk}, are grouped together during days when κ and l are implemented. The individual standard deviations (J − 1) of each df are calculated based on repetitions of j as follows:

$$\sigma_{1V\kappa} = \sqrt{\left(\frac{1}{J-1}\right)\sum_{j=1}^{J}\left(Y_{l\kappa j} - \bar{Y}_{l\kappa}\right)^2} \quad where \ \bar{Y}_{l\kappa} = \left(\frac{1}{J}\right)\sum_{j=1}^{J} Y_{l\kappa j} \qquad [3.53]$$

Level II: the standard deviation of reproducibility (l) can be calculated based on data. The level II difference (σ_{2l}) is implemented on (l). The individual standard deviations $df(\kappa - 1)$ are calculated based on daily means of κ as follows:

$$\sigma_{2l} = \sqrt{\left(\frac{1}{K-1}\right)\sum_{\kappa=1}^{K}\left(Y_{l\kappa.} - \bar{Y}_{l..}\right)^2} \quad where \ \bar{Y}_{l..} = \left(\frac{1}{K}\right)\sum_{\kappa=1}^{K} Y_{l\kappa..} \qquad [3.54]$$

Level III: a unique global difference can be calculated based on the L-execution = (L-run) of the medium. A difference of level 3(L − 1) df is calculated based on the L-run of the medium as follows:

$$\sigma_3 = \sqrt{\left(\frac{1}{L-1}\right)\sum_{l=1}^{L}\left(\bar{Y}_{l..} - \bar{Y}_{...}\right)^2} \quad where \ \bar{Y}_{..} = \left(\frac{1}{L}\right)\sum_{l=1}^{L} \bar{Y}_{l..} \qquad [3.55]$$

The standard deviation of uncertainty for a single measurement on a test element is written as follows:

$$\sigma_R = \sqrt{\sigma_{active}^2 + \sigma_{days}^2 + \sigma_1^2} = \sqrt{\sigma_3^2 + \left(\frac{K-1}{K}\right)\sigma_2^2 + \left(\frac{J-1}{J}\right)\sigma_1^2} \quad [3.56]$$

The values σ_1 and σ_2 are:

$$\sigma_1 = \sqrt{\frac{\sum_{l=1}^{L}\sum_{\kappa=1}^{K}\sigma_{1l\kappa}^2}{LK}} \quad and \quad \sigma_2 = \sqrt{\frac{\sum_{l=1}^{L}\sigma_{2l}^2}{L}} \quad [3.57]$$

c) Problem of estimation of degrees of freedom: if the df are necessary for uncertainty, the formula above cannot be used directly in terms of the standard deviation: σ_1, σ_2, and σ_3, we propose:

$$\sigma_{reported\_value} = \sqrt{\frac{\sigma_3^2}{p} + \left(\frac{K-M}{PMK}\right)\sigma_2^2 + \left(\frac{J-N}{PMNJ}\right)\sigma_1^2} \quad [3.58]$$

The sensitivity coefficients are:

$$\left\{\psi_1 = \sqrt{\left(\frac{J-N}{PMNJ}\right)};\ \psi_2 = \sqrt{\left(\frac{K-\mu}{PMK}\right)};\ \psi_3 = \sqrt{\left(\frac{1}{P}\right)}\right\} \quad [3.59]$$

The sensitivity coefficients specific for three uncertainty components are indicated in the Table 3.5 for the selections of *N, M P.* J must be ≥ N and k must be ≥M. Furthermore, the following constraints must be observed:

Short-term number, *N*	Daily number, *M*	Number (r & r), *P*	Short-term sensitivity coefficient, a_1	Daily sensitivity coefficient, a_2	(r & r) sensitivity coefficient, a_3
1	*1*	*1*	$\sqrt{\frac{J-1}{J}}$	$\sqrt{\frac{K-1}{K}}$	*1*
N	*1*	*1*	$\sqrt{\frac{J-N}{NJ}}$	$\sqrt{\frac{K-1}{K}}$	*1*
N	*M*	*1*	$\sqrt{\frac{J-N}{MNJ}}$	$\sqrt{\frac{K-M}{MK}}$	*1*
N	*M*	*P*	$\sqrt{\frac{J-N}{PMNJ}}$	$\sqrt{\frac{K-M}{MPK}}$	$\sqrt{\frac{1}{P}}$

Source: *Engineering Statistics Handbook* (USA).

Table 3.5. *Sensitivity coefficients for three uncertainty components*

Let us calculate a standard deviation for each component of type B uncertainty. Next, let us combine type A standard deviation and type B standard deviation with a standard uncertainty for the result signaled with the help of sensitivity factors.

3.5.2.2.3. Sensitivity factors

A final value involves more than just quantity. We will present the steps to be followed for an uncertainty evaluation including secondary quantities:

1) Write the equation showing the relationship between quantities, i.e. the equation of the propagation of errors and carrying out a preliminary evaluation.

2) If the measurement can be taken independently of the number of secondary quantities in individual repetitions, treat the uncertainty evaluation, making sure to evaluate all the sources of random error in the process.

3) If the measurements cannot be reproduced directly, treat each measure by

i) calculating the standard deviation for each quantity of measurement;

ii) combining the standard deviations for individual quantities and the standard deviations for the result which comes from the propagation of errors.

4) Calculate a standard deviation for each component of type B uncertainty.

5) Combine the standard deviations of *type A* uncertainty and *type B* uncertainty with a standard uncertainty (u) for the result signaled.

6) Calculate an expanded uncertainty (U).

7) Compare uncertainty derived from the propagation of errors and the uncertainty obtained using techniques of data analysis.

3.5.2.2.4. Evaluations of type B uncertainty

Evaluations of type B uncertainty apply to bias and to error: the characteristic factor is that the calculation of the uncertainty component is not based on a statistical data analysis. The distinction to keep in mind in terms of random error and bias is that random errors cannot be corrected and that bias can be corrected or eliminated using the result.

There are a few cases of sources of uncertainty which lead to *type B* evaluations:

– norms of reference calibrated by another laboratory;

– physical constants used in the calculation of the final value;

– environmental effects which cannot be selected;

– lack of resolution of the instrument.

Sources of uncertainty in other processes: sources of uncertainty such as calibration relationships for reference measurement standards or relationships said to be uncertainties for physical constants do not pose any difficulty in the analysis. Uncertainty will generally be signaled as expanded uncertainty *U*, which is converted into standard uncertainty, $u = U/\kappa$. If the coefficient κ is neither known nor well documented, it is recommended that it be considered to be $\kappa = 2$.

Sources of uncertainty which are for the measurement process: sources of uncertainty which are local for the measurement procedure, but which cannot adequately be selected to permit a statistical analysis demand type B evaluations. According to the GUM, it is convention to *attribute infinite df* to standard deviations derived in this way.

Standard deviations of presumed distributions: estimations of the difficulty of obtaining reliable uncertainty: the methods attempt to avoid the difficulty of sources of error authorization for which reliable estimations of uncertainty do not exist. The methods are based on hypotheses which can be valid or invalid and demand the experimenter's consideration of the effect of the hypotheses on final uncertainty.

Estimations of the difficulty of obtaining reliable uncertainty: the approach is to consider all errors and bias, for the situation in hand to make an impression in the form of a known statistical distribution. So, the difference is calculated based on known (or assumed) characteristics of the well-known distribution. The distributions which can be considered are *uniform, triangular*, or *normal* (*Gaussian*).

Standard deviation of a uniform cumulative distribution: uniform cumulative distribution leads to the estimation of uncertainty, i.e. with a large difference. The calculation of the standard deviation is based on the hypothesis that the evaluation criteria, ±c, of the distribution are known. It is the hypothesis where all the effects on the final value, between − c and + c, are all susceptible to the particular source of uncertainty.

Table 3.6 summarizes the main laws of distribution [GRO 09, GRO 11]

Triangular distribution leads to an estimation of uncertainty, i.e. with a smaller difference than the uniform distribution. The calculation of the standard deviation is based on the hypothesis that the evaluation criteria, ±c, of the distribution are known and the mode of the triangular distribution occurs at zero (see Table 3.6). Normal distribution leads to the less conservative estimation of uncertainty. The calculation of the standard deviation is based on the hypothesis where the evaluation criteria, ±c, engulf 99.7% of the distribution.

Range 2 and mean $\mu = 0$ for each of the three distributions approached Measurement uncertainty published and reduced according to statistical distribution			
Statistical distribution laws appropriate in operation	Distribution graphs Graph showing the statistical distribution around the value x_i	Variance σ^2	Standard deviation σ
Normal distribution (Gauss–Laplace) If the uncertainty or the standard deviation are given in an expanded form, we divide uncertainty by factor k: 99.73% ($\kappa = 3$) $C = 3\sigma U(\cdot 1)$ = range/k	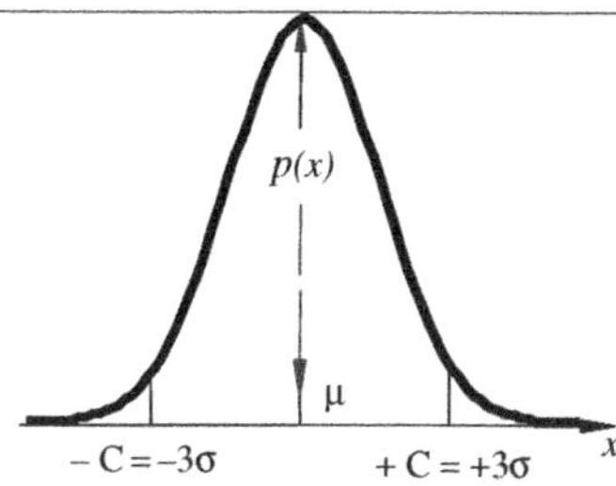	$\frac{C^2}{9}$	$\frac{C}{3}$
Uniform distribution or *rectangle* Used for statistical measurement standards (calibration) and numerical displays. The extreme values (c)+ and (c)– are given and probable = general case, given by default in cases of doubt = 'if need be' we divide by $\sqrt{3}$	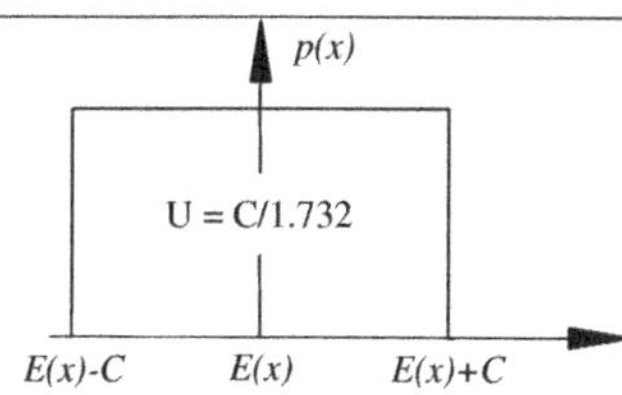	$\frac{C^2}{3}$	$\frac{C}{\sqrt{3}}$
Derivative of the function: *Arc sinus.* It is used for periodic disturbances to analyze the effect of influence values varying between two *extremums* in a sensitively sinusoidal way for example the local temperature of a workshop. If temperature variations $t°$ are indicated by ±C then $\{u = c/\sqrt{2}\}$	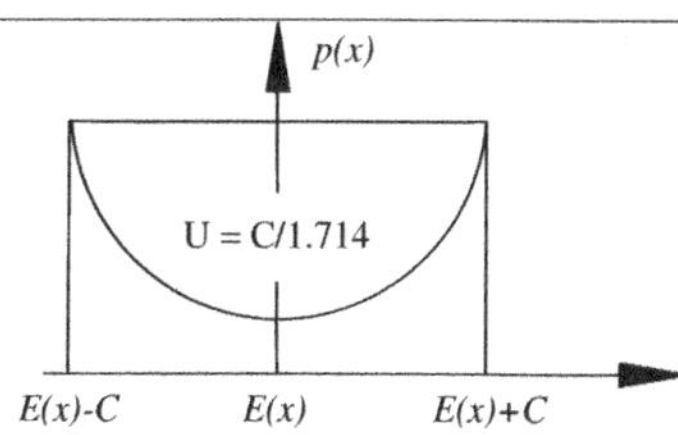	$\frac{C^2}{2}$	$\frac{C}{\sqrt{2}}$
Triangular distribution law Often used for parallelism. The measured values are generally closer to the "center" than the extremes. In the rarest of cases, we divide by $\sqrt{6}$	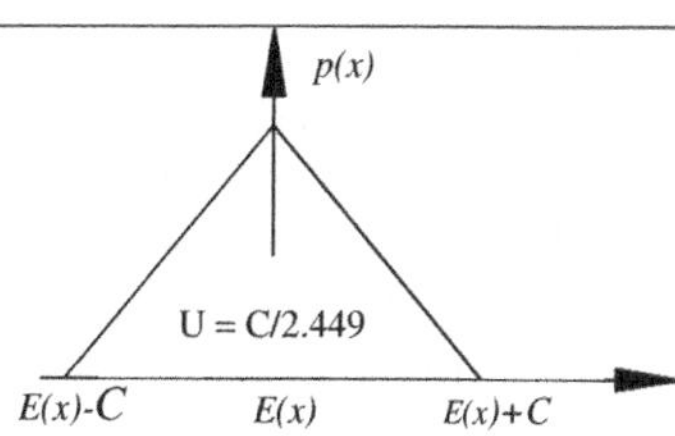	$\frac{C^2}{6}$	$\frac{C}{\sqrt{6}}$

Table 3.6. *Recapulative table of ordinary distributions for the calculation of type u(x) uncertainty*

In the context of the use of the Welch–Satterthwaite relation [BRO 60] with the above distributions, the number of *df* is supposed to be infinite (see [3.2]). In fact, the *type B* method would make use of the manufacturer's statistical data through a *devious* intervention of the singularity of the experiment, but not of the instrument whose manufacturer supplies us with the typical uncertainty which belongs to it. If we decided to try (during a measurement process) bringing in a correction, this correction x_i would, in fact, be unknown. We only know the limits between which it is would be situated, i.e. between the $C_{lower\ limit}$ value (Lower limit) and the $C_{upper\ limit}$ value (Upper limit). We know from the literature [ACN 84, GUI 00, PRI 96] that the correction value will be estimated using the expression of x_i:

$$x_i = \frac{1}{2 \times \left\{ C_{\text{lower limit}} - C_{\text{Upper limit}} \right\}} \quad [3.60]$$

For an initially uniform or rectangle distribution and from Table 3.6, the estimator of the corresponding variance is written as follows:

$$\sigma^2(x_i) = \left\{ \frac{1}{12\left(C_{\text{lower limit}} - C_{\text{Upper limit}}\right)^2} \right\} \quad [3.61]$$

We note that $IT = 2C_i$, IT being the difference between two lower and upper limits:

$$\sigma^2(x_i) = \left\{ \frac{1}{3\left(C_{\text{Upper limit}}\right)^2} \right\} \quad [3.62]$$

We notice that statistical expressions remain inescapable, even in the case of the type *B* method. If we use language and GPS representations [CHA 99]; we will be able to express type uncertainty on x_i. Table 3.6 allows a structured reading of the distribution of methods *A* or *B* [GUI 00, NIS 94]. All that is needed is to indicate that the distribution derived from *Arc sinus* is used to calculate uncertainty on values which surround the *true value*. It is used in this way in the case of the recommended temperature between the instrument and the part. According to obtained precision, it has been appropriate to ensure the *homogeneity* of the ambient parameters surrounding the part.

The thermal equilibrium must be ensured between the part being tested and the material used for this test. The time for stabilization is important to master during the manufacturing processes or during other thermal treatments. It works according to the volume of the part and its dynamics.

$f_{(\tau)}$ =
1
0.607
0.368
0.223
0.135
0.082
0.05
0.03
0.018
0.011
$6.738 \cdot 10^{-3}$

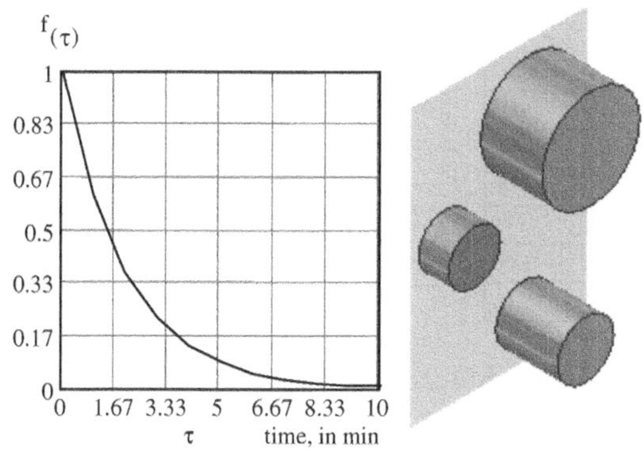

Figure 3.10. *Demonstration of stabilization time according to the surface function*

The stabilization time is around 30 times more for the large cylinder than the plaque and the small cylinder (made from the same material), in the same thermal conditions (initial and final). The $T_{stabilization}$ is a function of the following relationship:

$$Time_{stabilization} = Volume / Surface \qquad [3.63]$$

Here is an illustration of the influence of the exchange surface on the thermic stabilization time with the same initial and final conditions: temperature, 20°C; atmospheric pressure, 101,325 Pa; and hygrometry, 65%. The rate of hygrometry affects the dimensions of rubber, plastic, and granite parts. Vibrations can introduce recording errors. We must therefore ensure that there is an absence of any magnetic field. The weight of the part can also provoke an elastic deformation in the measuring apparatus. If the measurement is taken with a non-null force, the result must be corrected.

We are going to calculate the uncertainty type according to the veracity of the data, hence the interval $\{-C; +C\}$ of a uniform distribution. With ($x_{bar} = \mu$) mean of (observations) measurements and C_j, the correction of accuracy ($C_j = 0$ corresponds to the best state of our knowledge), the model of the measurement process is written as follows:

$$y = \bar{x} + \left\{ C_{jcorrected} \right\} \qquad [3.64]$$

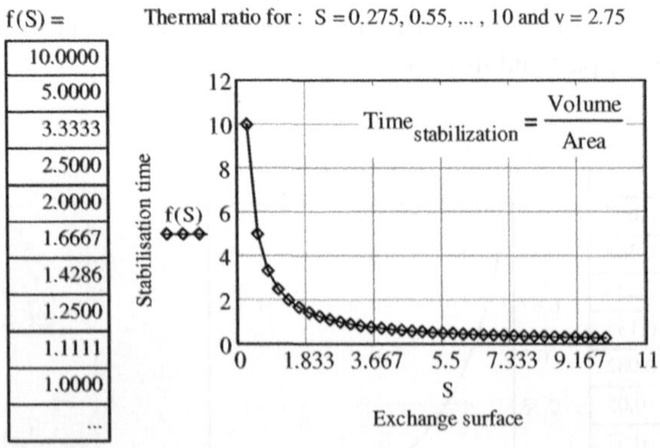

Figure 3.11. *Thermal ratio according to exchange surface*

From the law of propagation of uncertainties, we postulate, for unitary a_1 and a_2:

$$u_c^2(y) = a_1^2 \times u^2(\overline{x}) + a_2^2 \times u^2(C_i) = u^2(\overline{x}) + u^2(C_{jcorrected}) \qquad [3.65]$$

This relation also takes the simplified form in which s represents the experimental type of standard deviation of a series of n measurements. Thus, we propose the relation μ = 0.647741 and s = 0.000048, which are previously calculated. By applying [3.65] to our case, the propagation law of uncertainties $U = u^2(\mu)$ on the mean μ allows us to obtain, for example:

$$\text{For } s = 4.794673 \times 10^{-5} \text{ and } N = 10,\ u = \left(\frac{s^2}{N}\right) = 2.299 \times 10^{-10}$$

$u^2(C_j)$ will be estimated using the *type B* method. We consider the best correction value to be $u^2(C_j) = 0$ and we can see that the correction has the same probability of taking any value in {−10 μm and +10 μm}. The distribution is rectangular, as we have just seen, so the uncertainty type corresponds to this distribution within the limits{− C to C} and therefore u = (C/1.732) (see Table 3.6, second row):

For a correction,

$$C = 0.01,\ u = C/\sqrt{3} = 5.774 \times 10^{-3} \ \ So \ \ u^2 = 3.334 \times 10^{-5}$$

Let us calculate the uncertainty type (using its variance) as follows:

$$u^2(y) = u^2(\mu) + u^2(C_j)$$

$$u^2(y) = 0.001 \times 10^{-5}\,\text{mm}^2 + 3.334 \times 10^{-5}\,\text{mm}^2 = 3.335 \times 10^{-5}\,\text{mm}^2$$

Verification: $u = \sqrt{u_c(y)} = 5.775 \times 10^{-3}$

5.775×10^{-3} mm = 0.05775 m = 0.2273615 po = 0.2273615 µinch

$u_c^2(y) = 0.227\ 361\ 5$ µpo

The final result of the measurement will be written:

$Y = \mu \pm k \cdot u_c^2(y)$, for $\{k = 2\}$

$Y = 0.647\ 741$ po $\pm\ 2[0.2273615 \times 10^{-6}]$

$[0.647741] + [0.2273615 \times E{-6}]^2 \times 2 = 0.64774145$

$[0.647741] - [0.2273615 \times E{-6}]^2 \times 2 = 0.64774055$

We can clearly see that the error is totally insignificant to the point that we must look from the side of the seventh figure after the decimal places, which we don't count if we round the results up to three figures after the decimal places. We notice that the arithmetic mean has remained equal to 0.6547740. So, the quantified uncertainty will be:

$\{-C; +C\}$, i.e. $\{-0.64774055; +0.64774145\}$

3.6. Principal of uncertainty calculations of *type A* and *type B*

When we carry out measurement experiments on mechanical parts, we repeat them several times, in the same conditions, in the hope of obtaining a good trend. Unfortunately, there are always dispersions and it is for this reason that we settle for statistical modeling. Therefore, we consider a first order mean to begin with, which is the mean μ and then we calculate a variance σ^2 and a standard deviation σ on a sample of size n. We then proceed to the mathematical statistical approach. From the concrete measurement aiming for a true value x_i, we are taken to proceed with calculations of the mean of n samples, supposed values, with identical evidence. Systematic errors which come from this can be reduced, however, by applying corrections. A knowledge of the measurement procedure and of the fundamental principles which come from physics remains one of the best warrantors of the conduction of a metrology project. Various errors intervene during measurement, such as:

– temperature and pressure, and deformation of the mechanical part;

– accuracy of instruments and position of the entity being measured;

– disturbance of the value measured by the presence of the measurement instrument;

– error introduced by the measurement method itself or by the operator etc.

The rigorous work of a good metrologist remains that of thinking about errors which have not yet been identified. Those already identified will be the object of eventual corrections in order to compensate for them. Even if these corrections were to be judiciously brought forward, there is always a doubt about the value of the correction. Of the different corrections to be brought forward, three categories are particularly important:

1) The calibration corrections are determined and represented in the calibration certificates (CLAS – Calibration Laboratory Assessment Service, in Canada).

2) Environmental corrections make up the effect of influence values such as pressure and temperature. To take these into account, it is necessary to know their physical coefficients of sensitivity to different states.

3) Standardization corrections tend to bring back results in standard conditions, known as normal conditions (norms) of 20 °C ± 2 °C.

Measurand Y is linked by a function of measurable values $(X_0, X_1, \ldots, X_k)$. From [3.23] we note that $Y = F(X_1, X_2, \ldots, X_k)$. Each of these values $(X_0, X_1, \ldots, X_n)$ will be the object of compensating corrections for systematic errors. Respective corrections would take the following designations: $(C_1, C_2, \ldots, C_k)$. The application of global correction to y follows a Taylor development for the corrections to bring to $(C_0, C_1, \ldots, C_k)$ due to the fact that they are too small in front of the measured values $(X_0, X_1, \ldots, X_k)$:

$$C_y = \left(\frac{\partial F}{\partial x_1}\right)\cdot C_1 + \left(\frac{\partial F}{\partial x_2}\right)\cdot C_2 + \left(\frac{\partial F}{\partial x_3}\right)\cdot C_3 + \ldots + \left(\frac{\partial F}{\partial x_k}\right)\cdot C_k \qquad [3.66]$$

We must not expect the mathematical model to offer the integrality of the phenomena linked to the process. A model such as [3.66] constitutes a way of measuring easily in the mathematical formalism [GUI 00, MUL 81, NIS 94, PRI 96, TAY 05] which it suggests. However, there is a risk, in many respects, of offering a simplified vision of the metrology function.

For the final part of the analysis, we must say that the modeling of the measurement process arises as an important stage in view of estimating uncertainty. If $y = f(x)$ is the result of the measurement process, we propose:

$$f(x) = \bar{x} + \left\{C_{\text{standard}} + C_{\text{environment}}\right\} \qquad [3.67]$$

where:

$\overline{x}$ is the mean of the gross results of the measurement;

$S_{tandard}$ is correction due to calibration;

$C_{environment}$ is correction due to the environment.

In the case of multiplicative and additive corrections, the model will take the following mathematical form:

$$f(x) = \left(\sum_{i=1}^{k} C_{\text{standard}} \right) + \left(\prod_{p=1}^{m} C_{\text{environment}} \right) \times X \qquad [3.68]$$

where:

X is the rough measurement of the measuring instrument;

C_{addi} represents the additive corrections;

C_{mult} represents the multiplicative corrections.

An important factor persists, however, in being absent from our model. Unfortunately, we cannot quantify it as there is some *doubt.* In each of the factors of the function of the model, there is doubt surrounding the value attributed to the component. We use the arithmetic mean to represent the *best value*, but doubt still exists. There is also doubt at the level of the correction value. For calibration, correction is presumably indicated by its own uncertainty.

3.6.1. *Standard and expanded uncertainties*

The sensitivity coefficient and the standard deviations are combined with the sum of the square root to obtain a "standard uncertainty". Having been given components R, normalized uncertainty (u) is written as follows:

$$u = \sqrt{\sum_{i=1}^{R} \Psi_i^2 \times \sigma_i^2} \qquad [3.69]$$

If the object of the uncertainty declaration is to give a coverage with a raised level of confidence, expanded uncertainty is calculated using $U = \kappa \cdot u$. κ is chosen to be the $\alpha/2$ critical value of the Student's t-table with v as df. $\kappa = 2$ gets closer to a coverage of 95%. Expanded uncertainty is defined to give a high level of coverage for the true unknown value of the measurement $\text{Y} \rightarrow \{\text{Y} - \text{U} \leq \text{true value} \leq \text{Y} + \text{U}\}$.

The uncertainties are listed in Table 3.7 with their corresponding sensitivity Ψ_1, standard deviations and *df*.

Type A uncertainty components		
Coefficient of sensitivity	Standard deviation	Degrees of freedom (*df*)
Ψ_1	σ_1	ν_1
Ψ_2	σ_2	ν_2
Ψ_3	σ_3	ν_3
Type B uncertainty components		
Nominal test/nominal ref.	σ_4	ν_4

Table 3.7. *Type A and type B uncertainty components*

3.6.2. *Components of type A and type B uncertainties*

The sensibility coefficient shows the relation of the individual uncertainty component compared to the difference of the final value for the test element. The sensibility coefficient relates to that which is signaled and not to the estimation method of uncertainty components where uncertainty:

$$u = \sqrt{\sum_{i=1}^{R} \psi_i^2 \times s_i^2} \quad [3.70]$$

3.6.3. *Error on repeated measurements: composed uncertainty*

If we try, using a measurement, to obtain the true value x_o of a physical value, we would notice the measurements of this size successively being repeated, leading to different results $x_1, x_2, \ldots, x_n$. These are the results of *n* measurements carried out in identical conditions. Specialists call this doubt *uncertainties.* These uncertainties are numerically expressed using quantities which make us aware of the phenomenon of dispersion, either standard deviations $U(x_i)$ (*standard uncertainty*) or variances $U^2(x_i)$. The corrections figuring in the model whose value is unknown must appear. The reason for this is that corrections, even when null, have a variance and uncertainty type. Each of the quantities $x_1, x_2, \ldots, x_n$ is affected by an uncertainty $u(x_1)$, $u(x_2)$, $u(x_3)$, …, $u(x_n)$. These quantities are expressed under the form of the standard deviation and it is for this reason that we call them uncertainty types. The components combined with the uncertainty of the measurement results *y*, represented by $u_c(y)$, are calculated using the expression of estimated variance $u_c^2(y)$. Sometimes, we use the variance whose expression is given as follows:

$$u_c^2(y)=\sum_{i=1}^{k}\left(\frac{\partial F}{\partial x_i}\right)^2\cdot u^2(x_i)+2\cdot\sum_{i=1}^{k-1}\cdot\sum_{j=i+1}^{k}\left(\frac{\partial F}{\partial x_i}\right)\cdot\left(\frac{\partial F}{\partial x_j}\right)\cdot u(x_i,x_j) \qquad [3.71]$$

This equation is based on the approximation of the first order Taylor series, very small measurement values which are taken into consideration by the equation of uncertainty error propagation, where x_i and x_j are the estimators of X_i, X_j, and $u(x_i, x_j) = u(x_j, x_i)$ is the covariance estimated from x_j, x_i. The degree of correlation R between x_i and x_j is characterized using the estimator of the correlation coefficient $R(x_i, x_j)$, with respect to properties previously presented, and is written as follows:

$$R(x_i,x_j)=\frac{u(x_i,x_j)}{u(x_i)\cdot u(x_j)} \qquad [3.72]$$

$R\{x_i, x_j\}$ is often obtained by mathematical interpolation or by averaging the least squares method. $R\{x_i, x_j\}$ is included between $\{-1$ and $1\}$. This is a matter of course if x_i and x_j are independent so $R(x_i, x_j)$ is non-null, the equation $u_c^2(y)$ will become:

$$u_c^2(y)=\sum_{i=1}^{k}\left(\frac{\partial F}{\partial x_i}\right)^2\cdot u^2(x_i)+\ldots+$$
$$2\times\sum_{i=1}^{k-1}\cdot\sum_{j=i+1}^{k}\left(\frac{\partial F}{\partial x_i}\right)\times\left(\frac{\partial F}{\partial x_j}\right)\times u(x_j)\times u(x_j)\times R(x_i,x_j) \qquad [3.73]$$

The partial derivatives $\partial F/\partial x_i$ are evaluated at the point which translates associated standard uncertainty with the estimated values of the measurement x_i, and $u(x_i, x_j)$ which is, in fact, the covariance associated with x_i and x_j. The more we use the measured n values, the more we will be able to approach the *true value* x_o of the considered size. When we combine different *doubts*, we obtain a *type of law* named the law of uncertainty propagation. The classic form for an estimator y of the measurand Y comes from several measurable, observed quantities. In fact, the expression of combined uncertainty previously represented by $u_c^2(y)$, when x_i and x_j are independent and $R(x_i, x_j)$ is non-null but included between $\{-1$ and $1\}$, will take the formula which comes from the mathematical interpolation applied by $u_c^2(y)$. This final relation could seem a little complicated if one is not familiar with the mathematical support which it refers to.

Finally, expression [3.67] is generalized using this:

$$Y = f\left\{\begin{array}{l}\text{Unique and repeated observations,}\\ \text{additive and/or multiplicative corrections,}\\ \text{taking into consideration physical constants if they are mastered}\end{array}\right\} \qquad [3.74]$$

$y = f(x)$ is the result of the measurement process. In this expression, a simple measurement instrument will be used. A calibration correction is applied on this and Co_e is another environmental correction. As μ is the arithmetic mean, considering that the covariances are all null, we will calculate from this the composed uncertainty of $y = f(x)$ which will take the form:

$$u_c^2(y) = u_c^2(\mu) + u_c^2(Co_{environment}) + u_c^2(Co_{additives}) + u_c^2\left(Co_{Ctes\_physical}\right) \qquad [3.75]$$

NOTE.– Physical constants are rarely corrected. We ignore the term $u_c^2\left(Co_{Ctes\_physical}\right)$. In terms of standard deviation, expression [3.75] is written as:

$$u_c(y) = \sqrt{u_c^2(\mu) + u_c^2(Co_{environment}) + u_c^2(Co_{additives})} \qquad [3.76]$$

where $u_c^2(x)$ represents the variance on the mean of gross measurements. It is sometimes estimated based on the sample made up of the series of measurements. In other cases, it is estimated based on tests having allowed the estimation of the repeatability variance of the measurement process. It is in this final case that metrology laboratory assistants are in the best position. The fact is that we have more data and information to estimate repeatability variance.

3.6.3.1. *Applications on the calculations of uncertainties from laboratories*

It is interesting to summarize the recommended approach in the following seven distinct steps:

– definition of the value to measure;

– analysis of the causes of error;

– research on the ways of compensating for errors (corrections if there are any);

– modeling of the measurement process;

– estimation of the uncertainty type based on the result;

– succinct expression of the result and its expanded uncertainty.

The expression of the systematic uncertainty is susceptible to errors of interpretation. *Type A* components are characterized by σ_i^2 variances or estimated standard deviations σ_i. *Type B* components should be characterized by $(u_j)^2$, thus being considered by approximations of corresponding variances from which we admit existence. The terms $(u_j)^2$ can be used as variances and u_j as standard deviations. It happens that we do not necessarily take account of some additive corrections and this is why we propose the approach, reduced to five succinct steps:

– definition of the measured value;

– measurement conditions (environment, atmosphere, etc.);

– analysis of the causes of error;

– determination of eventual corrections;

– modeling of the measurement procedure of the propagation law of uncertainties according to the international ISO/CEI guide.

Previously, we have seen how variance is reached. If there is no covariance, we will proceed to the estimation of the *repeatability* of the measurement procedure using the variance (s^2) formulated as follows:

$$\begin{aligned} u_c^2(y) &= \left(\frac{\partial F}{\partial x_1}\right)^2 u^2(x_1) + \left(\frac{\partial F}{\partial x_2}\right)^2 u^2(x_2) + \ldots + \left(\frac{\partial F}{\partial x_k}\right)^2 u^2(x_k) \\ &= \sum_{i=1}^{k}\left[\left|Co_i\right| \cdot u^2(x_i)\right]^2 \end{aligned} \quad [3.77]$$

To evaluate measurement uncertainty, the GUM analytical procedure has become an international norm. Is it the only method? No, there are other methods, particularly when it becomes complicated to formalize all of the factors leading to expanded uncertainty in one equation. In these cases, we settle for *interlaboratory* methods of comparison, using a lot of aptitude testing.

The GUM method is analytical. It is based on the writing of a physical model of the measurand noted: $y = f(x_1, x_2, x_3, \ldots, x_n)$. In this chapter, we have seen how the GUM indicates the way of evaluating uncertainty types of these input values ($x_1, x_2, x_3, \ldots, x_n$). It would be illogical, in an "uncertainty-mania" state of mind to apply the GUM method to all domains.

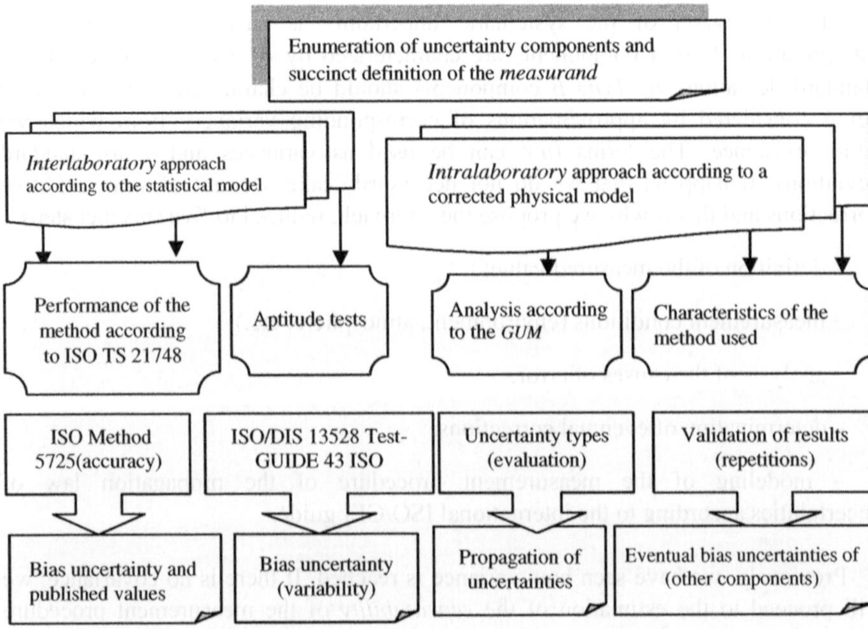

Figure 3.12. *Summary of the calculation methods of uncertainty [GRO 09, GRO 11]*

3.6.3.2. *Propagation of errors*

The approach (said to be descending) of uncertainty analysis which has survived until this point in the discussion has been what we call a high to low approach. Uncertainty components are estimated based on direct repetitions of the measurements of the result. This contrasts with the propagation of error approach. We take the simple example where we estimate the zone of a rectangle of repeated measurements of length and width. The following zone, *Zone* = (*length*) × (*width*), can be calculated based on each repetition. The difference of the signaled zone is directly estimated based on the repetitions of the zone. This approach (*descending* approach) presents the following advantages:

– appropriate treatment of covariances between the measurements of length and width;

– appropriate treatment of unexpected sources of error which would appear if measurements cover a range of conditions of functioning in a period lasting for a sufficient length of time.

For the independence of error propagation model, the propagation of errors combines the estimations of individual and auxiliary measurements. Formal propagation of the error approach involves calculating:

– the standard deviation of the width measurement;

– the two in a standard deviation of the zone with the help of approximation for double variable products (while ignoring a possible covariance between the length and the width).

$$S_{Area} = \sqrt{Width^2 \times S_{length}^2 + length^2 \times S_{Width}^2} \qquad [3.78]$$

Covariance terms can be difficult to estimate if the measurements are not done in pairs. Sometimes, these terms are deliberately omitted in the formula, if

1) the measurements of X, Z are independent, the term associated covariance is null;

2) the values of test elements of calibration have zero covariances.

The *descending* approach consists of estimating the uncertainty of direct repetitions of the result's measurement. That contrasts with a propagation of the error approach. Let's consider the example where we estimate the zone of a rectangle of repeated measurements of length and width. The zone is expressed as Zone = [*length*] × [*width*]. The difference of the zone signaled is directly estimated starting with the repetitions of the zone. This approach presents the following advantages:

– appropriate treatment of the covariances between the measurements of length and width;

– appropriate treatment of the unexpected sources of error emerging on the measurements of a sufficiently long period of time;

– independence of the propagation of the error model.

The formal propagation of the error approach involves calculating the difference based on the measurements of length and then the standard deviation of the width measurement. Then, it is appropriate to combine the two in standard deviation for the zone helping the approximation for the products with two variables

$$S_{area} = \sqrt{Width^2 \times S_{length}^2 + Length^2 \times S_{width}^2} \qquad [3.79]$$

3.6.3.2.1. Exact formula

Goodman Lee [GOO 60] proposed the derivative of an exact formula of difference between two products in 1960. Taking into account two random variables, x and y (corresponding to the width and the length of the approximate formula, respectively), the exact formula of the variance is:

$$V(\bar{x}\,\bar{y}) = \frac{\left(\begin{array}{l} X^2V(y)+Y^2V(x)+2XYE_{1,1}+\left(2X\dfrac{E_{1,2}}{n}\right)+\left(2Y\dfrac{E_{2,1}}{n}\right) \\ +\dfrac{V(x)V(y)}{n}+\dfrac{COV\{(\Delta x)^2,(\Delta y)^2\}-E_{1,1}^2}{n^2} \end{array}\right)}{n} \quad [3.80]$$

where:

$X = E(x)$ and Y correspond, respectively, to the width and the length;

$VA(x)$ is variance of x and $V(y)$ is the variance of $Y(= s^2)$ for the width and the length, respectively:

$$E_{i,j} = \{(\Delta x)^i,(\Delta y)^j\} \text{ with } \Delta x = (x - X) \text{ and } \Delta y = (y - Y)$$

$$COV\{\Delta x^2, \Delta y^2\} = E_{2,2} - VA(x)\times VA(y)$$

An estimation of the statistic is obtained by substituting the estimations of the sample for the corresponding values of the population. The approximate formula assumes that length and width are independent. The exact formula assumes that the length and the width are not independent. The propagation approach to error in structure reliability clearly presents the disadvantages. In an ideal case, the propagation of the error estimation above does not differ from the estimation carried out directly based on the measurements of regions. Sometimes, we cannot directly reproduce the measurement. It is therefore necessary to estimate its uncertainty via the propagation of error formulae [BRO 60]. The propagation of the formula of error for $Y = f(X, Z, \ldots)$, a function of one or several variables with the measurements, $X, Z, \ldots$, gives the estimation for the standard deviation of y

$$\sigma_y = \sqrt{\left(\frac{\partial Y}{\partial X}\right)^2 \sigma_x^2 + \left(\frac{\partial Y}{\partial Z}\right)^2 \sigma_z^2 + \ldots + \left(\frac{\partial Y}{\partial X}\right)\left(\frac{\partial Y}{\partial Z}\right)\sigma_{xz}^2 + \ldots} \quad [3.81]$$

where:

σ_x is measurement difference x;

σ_z is measurement difference z;

σ_y is measurement difference y;

σ_{xz} is the estimated covariance between measurements (X, Z);

$(\partial Y/\partial X)$ is the partial derivative of the Y function compared to X, etc.

Covariance terms can be difficult to estimate if the measurements are not carried out using pairs. Sometimes, these terms are omitted in the formula. Practically, the modalities of covariance should only be included in the calculation if they have been estimated with a sufficient amount of data. In Table 3.8, we present the important formulae used to calculate measurement uncertainty.

Recapulative table of calculations of composed measurement uncertainty, by propagation	
Formula of the calculation of result (y)	Calculation of composed uncertainty $U_{(y)}$, all correlations excluded: $y = f\{x_1 + x_2 + x_3 + ... + x_n\}$ Partial derivation prevents the calculation of composed measurement uncertainty. It is the law of the propagation of uncertainty
Sums and differences $y = \{x_1 + x_2 + ... + x_n\}$	$U_{(y)} = \sqrt{U_1^2 + U_2^2 + U_3^2 + ... + U_n^2}$
Products and quotients $y = \left\{x_1 \times \frac{x_2}{x_3} \times ... \times x_n\right\}$	$U_{(y)} = y \times \sqrt{\left(\frac{U_1^2}{x_1^2}\right) + \left(\frac{U_2^2}{x_2^2}\right) + \left(\frac{U_3^2}{x_3^2}\right) + ... + \left(\frac{U_n^2}{x_n^2}\right)}$
Exponents $y = \{x_1^a \times x_2^b \times ... \times x_n^i\}$	$U_{(y)} = y \times \sqrt{\frac{a^2 \times U_1^2(x_1)}{x_1^2} + \frac{b^2 \times U_2^2(x_2)}{x_2^2} + ... + \frac{i^2 \times U_n^2(x_n)}{x_n^2}}$

Table 3.8. *Calculation of composed uncertainty $U_{(y)}$*

We present two concrete examples to back up the above. To do this, we have chosen a simple case which comes up in metrology with a simple parabolic branch (see Figure 3.13) and a second case applied to fracture mechanics (fatigue) represented by Paris' law (see Chapter 1)

3.7. Study of the basics with the help of the GUMic software package: quasi-linear model

Roughness is calculated using the following relation:

$$R_a(f) = \frac{1}{8} \times \left(\frac{f^2}{R_\varepsilon}\right) \qquad [3.82]$$

where:

f is the feed = 0.15, 0.25,…, 1.5 mm/turn;

R_ε is the corner radius of the tool = 1.6 mm.

First case study: calculate the mean and the standard deviation of a sample of experimental data from data recorded on the surface indicator (our dimensional metrology laboratory [GRO 09, GRO 11]). With the help of GUMic software [GUM 08], we proceed to the appreciation of combined uncertainty using GUM methods and the Monte Carlo simulation, on a part rectified using conventional manufacturing (see equation [3.82]).

$R_a(f)$ = roughness resulting from manufacturing in μm;

R_ε is the corner dimension of the cutting tool) in μm;

n = lines (data) = 14 points of data in mm;

Mean = mean(data) = 0.063;

Standard deviation = SD(x) = $\sigma(x)\times\sqrt{n/(n-1)}$ = SD (data) = 0.053.

The uncertainty interval of X (or R_ε) leads to an uncertainty interval of Y (or R_a). What will be the probability associated with Y? As the uncertainty interval of X is quite small, we calculate Ra according to tool f in mm/turn and to the geometry of this tool (corner radius) R_ε in mm. Our results are:

Ra(f) =
$1.758\cdot10^{-3}$
$4.883\cdot10^{-3}$
$9.57\cdot10^{-3}$
0.016
0.024
0.033
0.044
0.056
0.071
0.086
0.103
0.122
0.142
0.164

$$\mathrm{Ra}(f) := \frac{f^2}{8\mathrm{x}(\text{ Corner radius of the tool })}$$

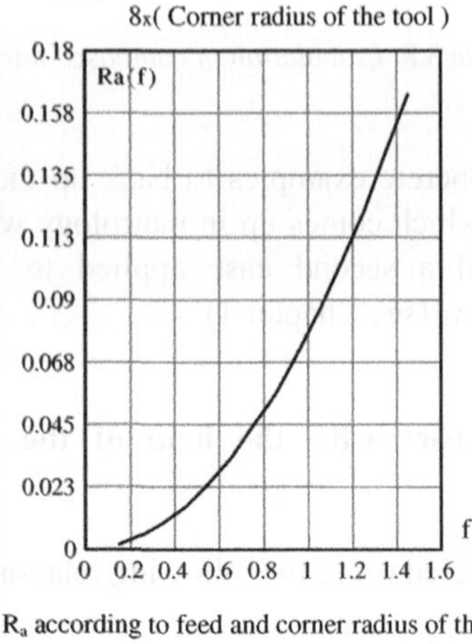

R_a according to feed and corner radius of the tool

Y is a function which is ready for linearization for this (small) interval, function [3.82] constitutes a parabolic branch $Y = X^2$ which we can assimilate to a straight line.

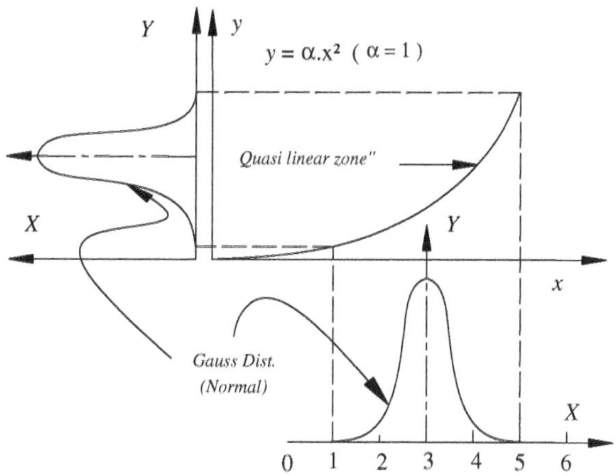

Figure 3.13. *Quasi-linear model for the calculation of uncertainty [GUM 08] of roughness*

Symbol value, X_i	Unit	Source of uncertainty	Probability law	Mean, $\overline{X}$	Standard deviation (X) or half range (U = R/2)	(n) sample size or reliability Δu/u in %
f	mm/ turn		Repeatability for $\overline{X}$	0.10	S = 0.001	n = 30
R_ε	mm		Repeatability for $\overline{X}$	1.20	S = 0.0025	n = 30

Table 3.9. *Statistics of experimental data of correction [GRO 09, GRO 11]*

Second case study: calculate the mean and the standard deviation of a sample of experimental data from data recorded on surface indicators. With the help of the GUM method of order 1 (k = 2 at the threshold of 95%) and that of MC,

1) calculate the composed uncertainty unit of size: $U_c(Y)$;

2) calculate the uncertainty type, unit of size: $U(x_i)$;

3) calculate the coefficient of sensitivity $\partial Y/\partial x_i$;

4) calculate the uncertainty, measurand unit (Y), $\frac{\partial Y}{\partial x_i}.\,u(x_i)$;

5) compare the results of the GUM method and that of MC simulation;

6) graphically trace the measurand Y distribution while comparing it to normal distribution.

Results of calculations and discussions

<table>
<tr><td colspan="4">Final result: $Y = R_a = (1.5 \times E{-}3 \pm 0.011) \times E{-}3$ µm with k = 2.05 at the threshold of 95%</td></tr>
<tr><td>Measurand, Y</td><td>GUM method of the first order</td><td>MC method Monte Carlo</td><td>Curves of the measurand</td></tr>
<tr><td>Mean</td><td>1.5 × E–6</td><td>1.5 × E–6</td><td rowspan="3">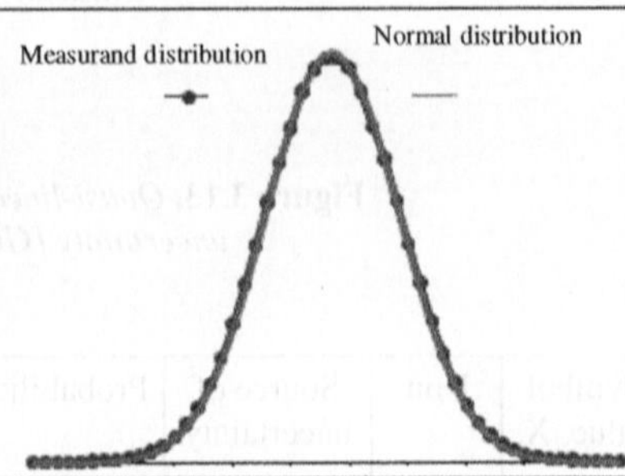
</td></tr>
<tr><td>Composed uncertainty, $U_c(Y)$</td><td>5.51 × E–6</td><td>5.51 × E–6</td></tr>
<tr><td>Number of df in operation, ν_{eff}</td><td>29</td><td>MC simulation (see Chapter 7, volume 2, reliability)</td></tr>
<tr><td colspan="4">The results are identical for the two methods therefore the measurand is quasi-linear on its uncertainty. This clearly validates the illustration of Figure 3.13.
The number of df is <30 so expansion factor $k = 2.05$ follows a Student's t distribution.</td></tr>
</table>

<table>
<tr><td>Uncertainty type. Unit of size, $U(x_i)$</td><td>Coefficient of sensitivity, $\partial Y/\partial x_i$</td><td>Uncertainty type unit measurand $(\partial Y/\partial x_i).\,u(x_i)$</td><td>Number of df in use, ν_{eff}</td></tr>
<tr><td>1.83 × E–4 mm/turn</td><td>0.0300 (µm/mm/trun)</td><td>5.48 × E–6 (µm)</td><td>29</td></tr>
<tr><td>4.65 × E–4 mm</td><td>1.25 × E–3 (µm/mm)</td><td>0.57 × E–6 (µm)</td><td>29</td></tr>
<tr><td colspan="4">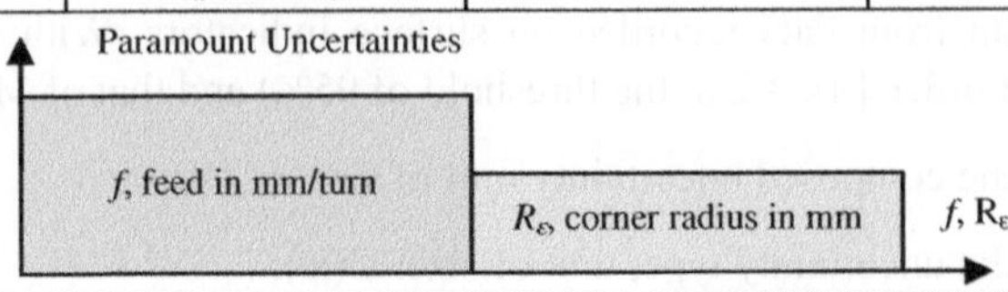
</td></tr>
</table>

Table 3.10. *Calculation of uncertainty in mechanical fabrication (roughness)*

DISCUSSION.– Due to validation concerns, we have used the GUMic program – GUM method and the MC method. The results are absolutely the same, to be noted: $u_c(y) = 5.51 \times E - 6$. The measurand, i.e. the value undergoing measurement, follows a normal distribution. The function (mean) coming from the first method is $Y = R_a = 1.5 \times E - 3 \pm 0.011\ \mu m$.

Third case study: it is dedicated to fracture. The third study case is dedicated to the mechanics of the fracture by fatigue the mean and the standard deviation of a sample of experimental data from recorded data [LAS 92] and [GRO 94, GRO 98] on the propagation of the fracture (a_0) of a structure soldered in a cross shape using four different soldering processes. This structure is submissive to a constant amplitude constraint $\Delta\sigma = 150$ MPa. Paris' law allows:

$$\frac{da}{dN} = C \times (\Delta K)^m \qquad [3.83]$$

where:

da/dN is the geometry relationship of the number of cycles applied, mm/cycles;

C and m are intrinsic parameters from the material (adimensional);

ΔK is the factor of constraint intensity in $\left(\text{MPa} \times \sqrt{\text{m}}\right)$.

Size, X_i	Unit	Source of uncertainty	Probability law	Mean, $\overline{X}$	Standard deviation, s(X)	Sample size, n
ΔK, *s.i.f*	$\text{MPa}\sqrt{m}$	Sampling	Repeatability for X	111.12	$S = 6.269589$	50
C		Literature	Linear regression	3.25	$S = 1.70$	50
m		Literature	Linear regression	5.60 × E–12	$S = 0.004$	50

Table 3.11. *Statistics of experimental data of fracture: Paris' law*

Results of calculations and discussions

<table>
<tr><td colspan="4">Final result: Y = da/dN =(3.3 ± 4.4) × E−6 (mm/cycles) with k = 2 at the threshold of 95%</td></tr>
<tr><td>Measurand, Y</td><td>Method of GUM order 1</td><td>MC method</td><td>Curves of the measurand, accuracy = 100%</td></tr>
<tr><td>Mean</td><td>3.25</td><td>3.25</td><td rowspan="3">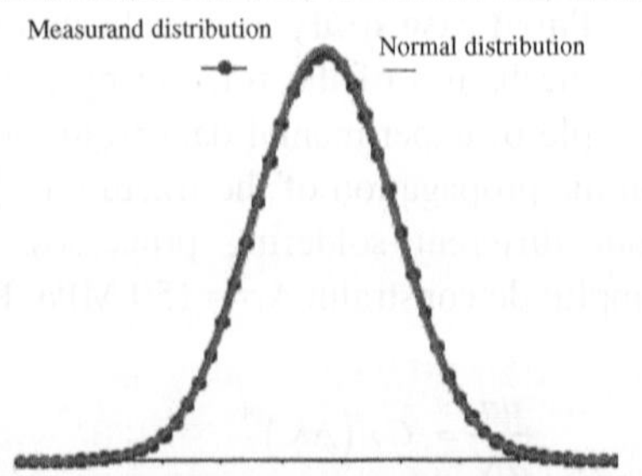
</td></tr>
<tr><td>Composed uncertainty $U_c(Y)$</td><td>1.70</td><td>1.70</td></tr>
<tr><td>Number of df in operation, ν_{eff}</td><td>48</td><td>MC simulation (see Chapter 7, volume 2)</td></tr>
<tr><td colspan="4">The results are identical for the two methods; therefore, the measurand is quasi-linear on its interval of uncertainty. This clearly validates the illustration of Figure 3.13. The number of df is <50 so the expansion factor k = 2 at the threshold of 95%.</td></tr>
</table>

<table>
<tr><td>Uncertainty type. Unit of size $U(x_i)$</td><td>Coefficient of sensitivity, $\partial Y/\partial x_i$</td><td>Uncertainty type unit measurand, $(\partial Y/\partial x_i)\times u(x_i)$</td><td>Number of df in operation, ν_{eff}</td></tr>
<tr><td>6.27 unit of ΔK</td><td>0.00 unit da/dN/unit ΔK</td><td>0.00 → unit da/dN</td><td>49</td></tr>
<tr><td>1.70 unit of C</td><td>1.00 unit da/dN/unit C</td><td>1.70 → unit da/dN</td><td>48</td></tr>
<tr><td>4.00E-3 unit of m</td><td>15.3 unit da/dN/unit m</td><td>0.06 → unit da/dN</td><td>48</td></tr>
<tr><td colspan="4">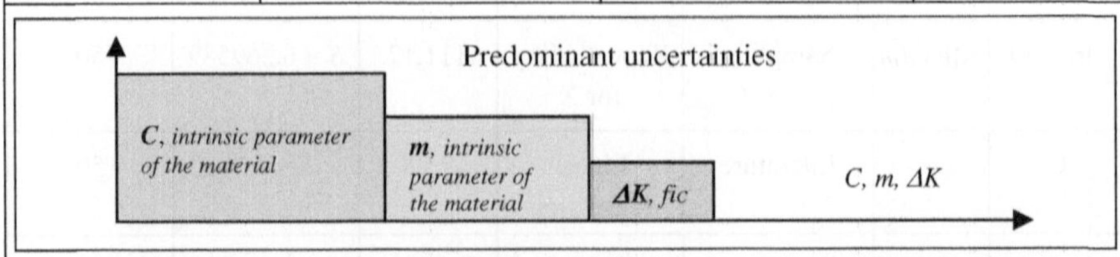
</td></tr>
</table>

Table 3.12. *Calculation of uncertainty in fracture mechanics: Paris' law*

DISCUSSION.– The two methods used are the GUM method and the MC method. The results to be noted are exactly the same: $u_c(y)$ = 1.70. The measurand, i.e. the value undergoing measurement follows a normal distribution. The function (mean) from the first method is $Y = da/dN$ = (3.3 ± 4.4) (mm/cycles) with k = 2 at the threshold of 95% with an accuracy of 100% for the result.

3.8. Conclusion

The causes of uncertainty are often dependent on the *repeatability* of measurements and the *environment.* Uncertainty characterizes the dispersion of values attributed to the *measurand. Expanded uncertainty* is a multiple of the *uncertainty type* which defines an interval in which we can hope to find a large number of values reasonably being attributed to the measurand. *Composed* uncertainty is the result of the propagation of all elementary components by means of linear mathematical approximation coming from a model of measurement. To carry out a *calculation of uncertainty*, it is indispensable to clearly state the model of measurement. The GUM reminds us that there is no correlation between *type A* and *type B* uncertainties. The MC simulation method is therefore a method which is indispensable in uncertainty estimation sometimes. The main aim of the calculation of uncertainty remains to be clear decision making. To do this, we need analytical results accompanied by a confidence interval.

3.9. Bibliography

[ACN 84] ACNOR (Association Canadienne de Normalisation) "Dessin Techniques et Principes Généraux", CAN3-B78.1-M83, Canada, 1984.

[AST 92] ASTM E691-92, Method, Standard Practice for Conducting an Interlaboratory Study to Determine the Precision of a Test Method, A Book of ASTM Standards, 10.05. PA 19428, 1992.

[BRO 60] BROWNLEE K.R., *Statistical Theory and Methodology in Science and Engineering*, John Wiley & Sons, New York, p. 236, 1960.

[CAM 74] CAMERON J.M., HAILES G.E., Designs for the Calibration of Small Groups of Standards in the Presence of Drift, Technical Note 844, U.S. Department of Trade, p. 31, 1974.

[CAM 77] CAMERON J.M., CROARKIN M.R.V., RAYBOL D.R.V., *Designs for the Calibration of Standards of Mass*, NBS Technical Note 952, U.S. Department of Trade, p. 58, 1977.

[CHA 99] CHARPENTIER J.A. DELOBEL J.P., LEROUS B., MURET C., TARAUS D., "Exploitation du concept GPS et de normalisation pour la Spécification Géométrique des produits", *CNAM*, Paris, 15 January 1999.

[CHU 62] CHURCHILL E., "Realistic evaluation of the precision and accuracy of instrument calibration systems", *Journal Research National Bureau of Standards-C. Engineering and Instrumentation*, vol. 67C, no. 2, pp. 161–187, 1962.

[CLA 00] CLAS., CALIBRATION LABORATORY ASSESSMENT SERVICE OF THE NATIONAL RESEARCH COUNCIL OF CANADA (NRC), CONAM Quantum Inspection and Testing, Burlington, Ontario, Canada, 2000.

[CLA 05] CALIBRATION LABORATORY ASSESSMENT SERVICE OF THE NATIONAL RESEARCH COUNCIL OF CANADA (NRC), Certificate of Calibration by Mitutoyo Calibration Laboratory according with ISO 10360-2, CAN3-85-Z299, 12th ed., 2001-12-15, 2005.

[DIX 51] DIXIS W.J., MASSEY F.J., *Introduction to Statistical Analysis*, McGraw Hill, NY, 1951

[FIC 90] FICKELSON M., "Test Quality for Construction,Materials and Structures", Proceedings of the International RILEM/ILAC Symposium, St Rémy-lès-Chevreuse, Edited by M.Fickelson, France, 1990.

[GER 91] GERALD J., MEEKER W.Q., *Statistical Intervals: A Guide for Practitioners*, John Wiley & Sons, New York, 1991.

[GOO 60] GOODMAN L., "On the exact variance of products", *Journal of the American Statistical Association*, vol. 55, pp. 708–713, 1960.

[GRO 94] GROUS A., Etude probabiliste du comportement des Matériaux et structure d'un joint en croix soudé, PhD thesis, UHA France, 1994.

[GRO 98] GROUS A., RECHO N., LASSEN T., LIEURADE H.P., "Caractéristiques mécaniques de fissuration et défaut initial dans les soudures d'angles en fonction du procédé de soudage", *Revue Mécanique Industrielle et Matériaux*, vol. 51, n° 1, Paris, April 1998.

[GRO 09] GROUS A., *Métrologie appliquée aux sciences et technologies*, collection capteurs et instrumentation dirigée par Dominique Placko. Édition Hermès, France 2009.

[GRO 11] GROUS A., *Applied Metrology for Manufacturing Engineering, Instrumentation and Measurement*, ISTE Ltd and Wiley, London and New York, NY, 2011.

[GUI 00] GUIDE., "Quantifier l'incertitude dans les mesures analytiques", *Eurochem/CITAC*, 2000.

[GUI 04] GUIDE pour la validation des méthodes d'essai et détermination de l'incertitude de mesure pour les laboratoires de la construction. Synthèse de la norme ISO/IEC 17025. Groupe de travail: Document N° 326.fw - Dr Nicolas Guscioni (auteur, responsable, CH). document n° 326, 31 pages. December 2004.

[GUM 08] GUMic Progiciel, Version 1.1. Login Entreprises, Poitiers, 2008.

[HOC 93] HOCKERSMITH T.E., KU H.H., "Uncertainties associated with proving ring calibrations", in KU H.H., *Precision Measurement and Calibration, Statistical Concepts and Procedures*, NBS Special Publication 300, vol. 1, pp. 257–263, 1993.

[HUN 86] HUNTER J.S., "The exponentially weighted moving average", *Journal of Quality Technology*, vol. 18, no. 4, pp. 203–207, 1986.

[ISO 97a] ISO 1109, Linear Calibration Using Reference Materials, ISO, 1997.

[ISO 97b] ISO 5725, Exactitude (justesse et fidélité) des résultats de la mesure, partie 2: méthode de base pour la répétabilité et la reproductibilité d'une méthode de mesure standard, ISO, Case postale 56, CH-1211 Genève 20, Suisse, 1997.

[JAE 84] JEAGER K.B., DAVIS R.S., *A Primer for Mass Metrology*, NBS Special Publication 700-1, p. 85, 1984.

[KU 66] KU H.H., "Notes on the use of propagation of error formulas", *National Bureau of Standards-C. Engineering and Instrumentation*, vol. 70C, no. 4, pp. 263–273, 1966.

[LAS 92] LASSEN T., Experimental Investigation and Probalistic Modelling of the Fatigue Crack Growth in Welded Joints, Agder College Faculty of Engineering, Grimstad, Norway, 1992.

[MUL 81] MULLER J.M., "Les incertitudes de mesure", in BUREAU DES LONGITUDES (eds), *Encyclopédie Scientifique de l'Univers, Tome 4, La Physique*, Gauthier-Villars, Paris, 1981.

[NIS 94] NIST., Guidline For Evaluating and Expression the Uncertainty of NIST Measurement Results based on the comprehensive ISO publication, Note 1297, (By TAYLOR B.N., KUYATT C.E.), 1994.

[NIS 97] NIST 1997, Engineering Statistics Handbook, F-Distribution, 1997. See also: http://www.nist.gov/itl/sed/gsg/strd.cfm.

[PRI 96] PRIEL M., *Métrologie dans l'entreprise, Mouvement Français pour la qualité*, AFNOR, Paris 1996.

[PHI 97] PHILLIPS S.D., EBERHARDT K.R., PARRY B., "Guidelines for Expressing the Uncertainty of Measurement Results Containing Uncorrected Bias", *NIST Journal of Research*, Institute of Standards and Technology, pp. 93–105, 1997, See also: http://www.docstoc.com/docs/7181405/Metrological-Timelines-in-Traceability

[REE 67] REEVE CHARLES P., "The calibration of angle blocks by comparison", *NBSIR*, 80-19767, Washington DC, 1967.

[TAY 94] TAYLOR B.N., KUYATT C.E., Guidelines for Evaluating and Expressing the Uncertainty of NIST Measurement Results Based on the Comprehensive International publication (ISO) Guide to the Expression of Uncertainty in Measurement, Note 1297, 1994.

[TAY 00] TAYLOR J., *Incertitudes et analyse des erreurs dans les mesures physiques*, Dunod, Paris , France, 2000.

[TUR 87] TURGEL R., VECCHIA D., "Precision calibration of phase meters", *IEEE Transactions on Instrumentation and Measurement*, vol. IM-36, no. 4, pp. 918–922, 1987.

[VIM 93] VIM, International Vocabulary of Basic and General Terms in Metrology, 2nd ed., ISO, Case postale 56 CH-1211, Geneva, Switzerland, p. 59, 1993.

Glossary

Abrasion resistance

Hard materials also show good abrasion resistance; in other words, they are not easily worn down by friction. In practice they are harder to grind down.

Acceptable risk

"*Acceptable risk*" describes the structural and non-structural measures to be put in place to reduce probable damage to a reference level. A risk scale is often associated with dangers to classify them in order of seriousness.

Availability

Is a (dimensionless) attribute of **dependability**. It is the capacity of a system to properly deliver the service (quality) when the user has need of it. Availability is a unitless measurement; it corresponds to the ratio of uptime to total execution time of the system.

Chance

Imaginary cause of something that occurs for no apparent or explicable reason (dictionary definition).

Conditional probability

(Bayesian) probability of a consequence when the causal event will definitely occur. If we suppose that a fracture has reached the limit suggested by a pre-established hypothesis, the probability of cracking is a conditional probability.

Corrosion resistance

Denotes the ability of a material to withstand damage from the effects of the chemical reaction of oxygen with the metal. A ferrous metal which is resistant to corrosion does not rust.

Dependability

The property which enables users to justifiably place their faith in the service provided to them: *reliability, availability, safety, maintainability,* and *security*.

Dilation and contraction

When a material is heated, it expands slightly: this is called this dilation. Conversely, if it shrinks (cold), this is contraction. The level of dilation and contraction of a metal affects its weldability. The more the metal is expanded or contracted, the greater the risk of *cracks* or deformations appearing.

Distribution function

Integral function of the probability density (or cumulative probability function), calculated in order of ascending values of the random variable. It expresses the probability of the random variable assuming a value less than or equal to a given value.

Ductility

Represents the ability of a metal to be deformed without breaking. It can be stretched, elongated, or subjected to torsion forces. Ductile materials are difficult to break because the cracks or defects created by a deformation do not easily propagate.

Elasticity

The ability of a material to return to its original form after a deformation.

Failure

Alteration or suspension of the ability of a system to perform its required function(s) to the levels of performance defined in the technical specifications.

Failure Mode, Effects, and Criticality Analysis (FMECA)

FMECA is a method for systematic risk analysis of the causes and effects of failures that might affect the components of a system. FMECA analyzes the seriousness of each type of failure. It enables us to evaluate of the impact of such failures on the reliability and safety of the system.

Fault tree

A *Logical* diagram using a tree structure to represent the causes of failures and their combinations leading to a **feared state** (Bayes). Fault trees enable us to calculate the **unavailability** or the **reliability** of the system model.

Fault (error) resistance

Is implemented to detect and handle errors.

Frailty

Frailty describes the characteristic of a metal that breaks easily on impact or from a deformation. It deforms little or not at all, and is easily broken.

Hardness

The ability of a body to resist penetration from another harder body. It is also characterized by its scratch resistance.

Hazard

Describes any event, unpredictable phenomenon, or human activity which would result in the loss of human lives, or damages to commodities or the environment.

Heat-affected zone (HAZ)

Martensitic coarse grain heat affected zone appears in welded joints. The heat affected zone is the area which has had its microstructure and properties altered by welding or heat intensive cutting operations.

Maintainability

One of the aspects of **dependability**. The maintainability of a system expresses its capacity for repair and evolution, with maintenance supposedly completed under certain conditions with prescribed procedures and means.

Malleability

A characteristic which allows the metal to be molded. It is the relative resistance of a metal subjected to compression forces. The malleability of a material increases with increasing temperature.

Markov chains

Used to evaluate the *dependability* of systems in a quantitative manner, this technique is based on the hypothesis that failure and repair rates are constant and that the stochastic process modeling its behavior is Markovian (a memoryless process). When the space of potential states of the system, is a discreat set the Markovian process is called a Markov chain.

Materialized measure

Measuring instrument which replicates or permanently provides different types of values during use, each with an assigned value.

Measurand

A value to be measured.

Measuring accuracy

Proximity between a measured value and the true value of a measurand.

NOTE:

– Measuring accuracy is not a value and is not expressed numerically. A measurement is sometimes considered *accurate* if it offers a *smaller uncertainty*.

– Although linked to the concepts of *correctness* and *fidelity*, it is better not to use the term "measuring accuracy" for measuring *correctness* or the term *measuring fidelity* for *measuring accuracy*.

– Measuring accuracy is occasionally associated with the proximity between the measured values attributed to the measurand.

Measuring instrument, measuring apparatus

Usually a device used for making *measurements*, on its own or possibly in conjunction with other devices.

Measuring repeatability

This is the measuring fidelity according to a set of repeatability conditions.

Measuring reproducibility[1]

This is the measuring fidelity according to a set of *reproducibility* conditions.

Measuring uncertainty

Non-negative parameter which characterizes the dispersion of values attributed to a measurand, arising from information used according to the method (e.g. A or B of the GUM).

Metrology

The science of measurements and its different applications, which encompasses all theoretical and practical aspects of measuring, regardless of the uncertainty of the measurement or the domain to which it relates.

Preliminary Hazard Analysis (PHA)

A method for identifying and evaluating hazards, their causes, their consequences, and the seriousness of these consequences. The aim of this analysis is to determine the appropriate methods and corrective actions to eliminate or control dangerous situations or potential accidents.

Probability

Statistical concept which can either express a degree of confidence or a measurement of uncertainty (subjective probability) or be taken as the limit of a relative frequency in an infinite series (statistical probability).

---

1 For statistical terminology refer to ISO 5725-1:1994 and ISO 5725-2:1994.

Probability density (or distribution function)

Function describing the *relative likelihood* of a random variable assuming a particular value. It assigns a probability to each value of a random variable.

Random

Process in which the result varies even if the input data set remains identical (a protocol leads to different results).

Reliability

The reliability of a system (work) is its aptitude to meet its design objectives over a specified period of time, in the environmental conditions to which it is subject to. Reliability is based on the probabilities used to evaluate it.

Reliability is one of the aspects of *dependability*. It corresponds to the continuity of service that the system must provide to its users, with the system being considered as irreparable. Any accidental failure is taken into account, regardless of its severity. Reliability measures the rate of failure, the inverse of the MTTF (*Mean Time-to-Failure*).

Risk

Risk is "a more or less predictable potential danger", or in other words a drawback which is more or less probable to which we are exposed. The scientific definition of risk involves an aspect of hazard and an aspect of loss, both expressed as probabilities.

Risk analysis

A system inevitably contains design errors, regardless of the amount of validation work done. The "zero error" criterion is not a realistic goal, in view of the development costs this would entail. Thus, it is important for the so-called critical systems to evaluate the risks for the users of these systems:

– *FMECA*: *Failure Mode, Effects and Criticality Analysis*

– *SEEA*: *Software Error Effects Analysis*

– *PHA*: *Preliminary Hazard Analysis*

Risk assessment

Procedure to determine the probability of a hazard *occurring* and its possible consequences.

Safety case

This approach is mainly used in the oil, nuclear, and rail transport sectors. In practice, this procedure facilitates the monitoring of studies.

Software reliability

Software inevitably contains *design errors*, no matter how strict the rules of design and validation. The ability of a software suite to provide acceptable service in spite of its residual errors defines its ***reliability***.

Security

We distinguish between safety and security. Thus

– Safety guards against catastrophic failures, for which the consequences are unacceptable in relation to the **risk**.

– Security relates to the prevention of unauthorized access to information.

Stress intensity factor (s.i.f., ΔK)

The stress intensity factor, K, is largely used in fracture mechanics (Irwin) to predict the stress state near the tip of crack usually as a the result of residual stress. Note that the magnitude of the stress intensity factor depends on the shape of geometry, the location and the size of the crack.

Tenacity

The ability of materials to resist shock without breaking or chipping.

Trend test

Is used in **reliability** to obtain indicators of reliability, from data on failures, and determine fluctuations in reliability over time.

Undesirable event

An event that should not occur or which should be improbable in view of the objectives in terms of *dependability*.

Unit (measuring)

A *Real scalar* value defined and adopted by convention, to which we can compare any other similar value to express the ratio between the two values as a number.

Value

Property of a phenomenon, (body, length, weight) expressed quantitatively by a number and a reference.

Abbreviations

CITAC	Cooperation on International Traceability in Analytical Chemistry
CSA	Canadian Standardisation Association
EA	European Cooperation for Accreditation
Eurachem	Focus for Analytical Chemistry in Europe
EUROLAB	European Federation of National Associations of Measurements, Testing and Analytical Laboratories
GUM	Guide to the expression of Uncertainty in Measurement (reference document recognized by the CSA, EUROLAB, Eurachem and EA)
IEC	International Electrotechnical Commission
ILAC	International Laboratory Accreditation Co-operation
ISO	International Organization for Standardization
S (or σ)	Standard deviation
SAS	*Service d'accréditation suisse* (Swiss accreditation service)
U	Uncertainty
VIM	International Vocabulary of basic and general terms in Metrology

Index

R

S

T

U, V, W

Printed and bound by CPI Group (UK) Ltd, Croydon, CR0 4YY

13/09/2026

14961075-0001